Götz Lehmann

Relativistic solitary structures in laser-plasma-interaction

Götz Lehmann

Relativistic solitary structures in laser-plasma-interaction

Models, Dynamics and Stability

Südwestdeutscher Verlag für Hochschulschriften

Impressum/Imprint (nur für Deutschland/ only for Germany)
Bibliografische Information der Deutschen Nationalbibliothek: Die Deutsche Nationalbibliothek
verzeichnet diese Publikation in der Deutschen Nationalbibliografie; detaillierte bibliografische
Daten sind im Internet über http://dnb.d-nb.de abrufbar.
Alle in diesem Buch genannten Marken und Produktnamen unterliegen warenzeichen-, marken-
oder patentrechtlichem Schutz bzw. sind Warenzeichen oder eingetragene Warenzeichen der
jeweiligen Inhaber. Die Wiedergabe von Marken, Produktnamen, Gebrauchsnamen,
Handelsnamen, Warenbezeichnungen u.s.w. in diesem Werk berechtigt auch ohne besondere
Kennzeichnung nicht zu der Annahme, dass solche Namen im Sinne der Warenzeichen- und
Markenschutzgesetzgebung als frei zu betrachten wären und daher von jedermann benutzt
werden dürften.

Verlag: Südwestdeutscher Verlag für Hochschulschriften Aktiengesellschaft & Co. KG
Dudweiler Landstr. 99, 66123 Saarbrücken, Deutschland
Telefon +49 681 37 20 271-1, Telefax +49 681 37 20 271-0, Email: info@svh-verlag.de
Zugl.: Düsseldorf, Heinrich-Heine-Universität, Diss., 2009

Herstellung in Deutschland:
Schaltungsdienst Lange o.H.G., Berlin
Books on Demand GmbH, Norderstedt
Reha GmbH, Saarbrücken
Amazon Distribution GmbH, Leipzig
ISBN: 978-3-8381-0454-6

Imprint (only for USA, GB)
Bibliographic information published by the Deutsche Nationalbibliothek: The Deutsche
Nationalbibliothek lists this publication in the Deutsche Nationalbibliografie; detailed
bibliographic data are available in the Internet at http://dnb.d-nb.de.
Any brand names and product names mentioned in this book are subject to trademark, brand or
patent protection and are trademarks or registered trademarks of their respective holders. The
use of brand names, product names, common names, trade names, product descriptions etc.
even without a particular marking in this works is in no way to be construed to mean that such
names may be regarded as unrestricted in respect of trademark and brand protection legislation
and could thus be used by anyone.

Publisher:
Südwestdeutscher Verlag für Hochschulschriften Aktiengesellschaft & Co. KG
Dudweiler Landstr. 99, 66123 Saarbrücken, Germany
Phone +49 681 37 20 271-1, Fax +49 681 37 20 271-0, Email: info@svh-verlag.de

Copyright © 2009 by the author and Südwestdeutscher Verlag für Hochschulschriften
Aktiengesellschaft & Co. KG and licensors
All rights reserved. Saarbrücken 2009

Printed in the U.S.A.
Printed in the U.K. by (see last page)
ISBN: 978-3-8381-0454-6

per aspera ad astra

Contents

1 Introduction   5

2 Physical model   13
    2.1 Maxwell-fluid equations 13
    2.2 Conserved quantities 17
    2.3 Numerical Algorithms 18

3 Techniques in stability analysis   23
    3.1 Idea of stability analysis 23
    3.2 Problems of linear stability analysis 26
    3.3 Linearized Maxwell-fluid equations 28

4 Stability and dynamics of relativistic 1D solitons   31
    4.1 1D model equations 32
    4.2 Solitons on the electron time scale 35
    4.3 Solitons on the ion time-scale 40
    4.4 Solitons in warm electron-ion plasma 55

5 Relativistic wave-breaking in cold plasma   65
    5.1 Laser wakefields for particle acceleration 65
    5.2 Wake-field excitation 70

Contents

    5.3   Wave-breaking calculations in Lagrangian coordinates 75

6 Two-dimensional dynamics of relativistic solitons   **93**

    6.1   Linearized 2D equations 94

    6.2   Transversal instability 97

    6.3   Nonlinear simulations 102

    6.4   Field structure 110

    6.5   Instabilities in 3D 121

7 Conclusion   **123**

A Appendix   **129**

    A.1   Stability of invariant sets 129

B Appendix   **131**

    B.1   Analytical stability criteria 131

Bibliography   **141**

1 Introduction

Since their invention in the 1960s, lasers are of continuously growing importance in physical research. Almost 50 years later the technology has been vastly improved in almost every way, from powerful lasers for industrial purposes to spectrally very narrowbanded continuous wave lasers for precise measurements of fundamental constants. Today a very large number of modern physical experiments would be impossible without lasers.

Nonlinear optical effects have been demonstrated shortly after the invention of lasers. This includes multiphoton ionization, modification of the refractive index of materials and disturbance of the Coulomb field of atoms. The first enhancements in terms of laser intensity were by methods such as Q-switching and mode locking. Intensities up to $10^{15}\,\mathrm{Wcm^{-2}}$ were feasible at the end of the 1960s.

A further increase in peak power depended on the possibility to amplify laser pulses with duration in the pico- or even the femto-second regime. In 1985 the chirped pulse amplification (CPA) was demonstrated [83], which lead to a very strong increase in obtainable peak power of lasers over the last 20 years.

Today peak intensities of $10^{21}\,\mathrm{Wcm^{-2}}$ are accessable on daily experimental basis [65, 70]. The next generation of lasers will reach up to $10^{26}\,\mathrm{Wcm^{-2}}$ [29]. This increase of up to ten orders of magnitude in peak power since the 1980s allows to access a great number of new nonlinear phenomena in the experiment. It is e.g. supposed that the interaction of such strong radiation with plasmas will provide a way to reach field

1. Introduction

intensities above $10^{28}\,\mathrm{Wcm^{-2}}$, which would exceed the Schwinger-field and lead to pair creation, a prediction made by QED theory [64].

One of today's most discussed application of such intense laser pulses are laser-plasma accelerators, which have been proposed as a new generation of particle accelerators [19, 84]. The accelerating electrical fields may be as large as 100 GV/m and more [27]. This is by many orders of magnitude larger than fields provided by conventional accelerator technology, which are limited to the order of roughly MV/m because of material breakdown. In plasma large field oscillations can be sustained, but the life-time of the oscillations may be limited due to wave-breaking. The plasma oscillations are driven by a relativistic laser pulse.

The interaction of high power lasers with plasma is said to be of *relativistic* nature. We suppose a linearly polarized laser and define the normalized amplitude a_0 of the laser vector potential as

$$a_0 = \frac{eA}{m_e c} = \sqrt{\frac{I\lambda^2}{1.4\cdot 10^{18}\frac{\mathrm{W}}{\mathrm{cm}^2}\mu\mathrm{m}^2}}\,, \qquad (1.1)$$

with laser peak intensity I, laser wave length λ, electron charge e, electron rest mass m_e, vacuum speed of light c and amplitude of the laser pulse vector potential A. The motion of charged particles in electromagnetic fields is determined by the Lorentz force. An electron irradiated by a laser pulse with $a_0 \ll 1$ performs harmonic oscillations transversely to the laser propagation. For $a_0 \gtrsim 1$ the force becomes nonlinear and the particle is accelerated in laser direction. The nonlinearity in the Lorentz force is introduced by the relativistic mass increase.

This nonlinearity is the source of many phenomena, such as laser pulse filamentation, relativistic plasma transparency, laser pulse self focussing, high order harmonic generation, excitation of nonlinear plasma waves and the generation of relativistic solitary structures [68].

Solitons or solitary waves are localized structures in nonlinear media. The interaction between solitons is particle-like, they emerge unchanged from an interaction. During the interaction however, their form may undergo considerable changes. Soliton solutions are known from many different areas of physics, the most prominent ones are fluid dynamics and fiber optics.

Solitons were predicted analytically in overdense plasma [42, 39, 23, 87], i.e. $\omega_0 < \omega_{pe}$, where ω_0 is the soliton frequency and $\omega_{pe} = (4\pi n e^2/m_e)^{1/2}$ is the electron plasma frequency for a plasma of density n. The solitons consist of trapped radiation and an associated plasma density variation, hence they have electromagnetic and electrostatic fields. In overdense plasma the pressure of the electromagnetic field is balanced by the excess pressure of the plasma from the outside.

A high power laser pulse propagating in an underdense plasma ($\omega_0 > \omega_{pe}$) will be influenced by nonlinearity, e.g. compressed. Stimulated Raman scattering and a Raman cascade causes a slow down of intensity spikes to $\omega_0 \approx \omega_{pe}$, which may lead to large amplitude relativistic electromagnetic solitons in an underdense plasma [55, 37, 56]. In addition to the nonlinearly shaped leading pulse, a laser beam propagating in underdense plasmas also creates slow, nearly standing narrow structures behind the leading edge. These processes are especially present in the ultra-short pulse regime [14].

Macroscopic evidence of soliton formation in multi-terawatt laser-plasma interaction has been reported from experiments [7, 8]. The bubble-like structures have been observed in numerical simulations, too [11, 81, 69, 66, 14, 24]. Within the solitons, ponderomotive pressure leads to a strong depression of the electron density. It is predicted that up to 40% of the laser energy can be trapped. The structures consisting of electron depressions and intense electromagnetic field concentrations are called slow solitons. Typical sizes of the spatial structures are of the order of the collisionless electron skin depth $d_e = c/\omega_{pe}$ of the surrounding plasma.

1. Introduction

The dynamics of soliton creation consists of two stages. *Pre-solitons* are created by the laser on the electron time-scale, which is in the order of $t \sim 1/\omega_{pe}$. The ions are to heavy to react to the oscillating fields on this time-scale. Pre-solitons can be either moving or standing structures. On the longer time-scale ($\sim 1/\omega_{pi}$), besides the electrons also the ions are pushed out of the density holes, and the solitons evolve into *post-solitons* [69]. The ion dynamics is responsible for a slowly expanding plasma cavity [53]. The expansion of the post-solitons under the push of the electromagnetic radiation (being trapped inside) has been analyzed within the snowplow approximation [69, 12]. Particle-In-Cell (PIC) simulations show merging (and not elastic interaction, as would be expected for true solitons) of post-solitons. A quite good agreement between experiment and PIC simulation occurs. Acceleration of solitons towards lower plasma densities has been observed [81] in agreement with theoretical expectations.

In analytical models for relativistic solitons stationary solutions are supposed. Most of the work on soliton solutions is for one-dimensional (1D) geometry and circular polarization [25, 42, 25]. The 1D geometry is a simplification which assumes that all quantities only depend on one spatial coordinate along the propagation direction. Within the cold relativistic hydrodynamic approximation the properties of solitons have been cataloged with respect to the number p of zeros of the vector potential, the velocity V, and the frequency ratio $\omega_0\sqrt{1 - V^2/c^2}/\omega_{pe}$ [$\equiv \omega\sqrt{1-V^2}$ in non-dimensional form]. Solitons do exist for $\omega^2(1-V^2) < 1$. Single-humped ($p = 0$) solitons have been found [25], e.g. for $V = 0$ when the ion response is neglected. Sub-cycle solitons ($p = 1, 2, \dots$) do exist for finite velocities V with a discrete ω-spectrum. On the ion time-scale, solitons do exist only above a certain threshold velocity V.

Circular polarized standing solitons in warm plasma were reported in Refs. [57, 58]. The solitons are derived from solutions of the relativistic Vlasov equation under the assumption of an isothermal plasma. The finite temperature introduces thermal pressure which is able to balance the radiation pressure of the soliton fields. This additional pressure is the reason for the existence of standing solitons in electron-ion plasma.

Linear polarized soliton solutions are only known on the electron time-scale and in the limit of weak plasma density response [32, 33].

The stability properties of solitons with respect to initial perturbations allow to draw conclusions about the life-time of such structures and the nonlinear evolution of the perturbation. Life-time and structure of the nonlinear state are important for possible experimental observation. In various publications the *1D stability* of solitons was investigated [14, 26, 72, 73, 58, 32, 33, 63, 76].

All general stability investigations use numerical methods, since only in limiting cases analytical expressions for the solitons are available. Usually the solitons are solutions to a system of coupled ordinary differential equations for the potentials **A** and ϕ. All other quantities like plasma density n and generalized plasma momentum **P** can then be calculated from **A** and ϕ. Previous investigations about stability of relativistic solitons were based on nonlinear simulations of the relativistic Maxwell-fluid equations dealing with soliton evolution. However, it is complicated to safely distinguish between a physical and a numerical instability in results from nonlinear simulations. Determination of the most unstable mode and the associated growth rate is usually not feasible by this method.

The development of an efficient numerical method to determine stability properties of different solitons is one focus of this work. It will be based on the linearization of a perturbation with respect to the unperturbed soliton. The most unstable mode and its growth rate will be determined by this method.

This stability analysis technique will then be applied to study the stability of solitons in different geometries. First we will focus on *longitudinal stability*, checking the stability of solitons which are perturbed by a small amount in propagation direction. We will study pre-solitons and post-solitons in cold and in warm plasma. The transi-

1. Introduction

tion dynamics of a pre-soliton into a post-soliton will be demonstrated. All this can be treated within the 1D framework.

The nonlinear stage of an unstable 1D soliton will result in the excitation of an electrostatic plasma wave behind the soliton. In general, it is possible for a high intensity laser pulse to excite plasma waves and by this transfer energy into the plasma. The excitation of plasma oscillations by relativistically intense laser pulses is a basic technique for laser-plasma based particle accelerators [31, 19, 84]. Experiments demonstrated the acceleration of electrons up to a few GeV energy [51], yet beam quality and energy spread are still problematic [28, 2, 41, 13, 61, 35, 44]. The exited oscillations may be very large, up to 100 GV/m and more [27].

The stability of the wake-field is of central interest for the laser-plasma accelerator scheme. Wave-breaking can limit the life-time of these fields. Analytical wave-breaking analysis of electrostatic plasma waves goes back to a paper from Dawson [19]. He studied plasma waves in the nonrelativistic case with homogeneous background density and found a critical threshold amplitude below which oscillations are stable. When the oscillation amplitude is larger, a multistream-flow sets in within the first oscillation and the wave breaks. In Ref. [21] a nonlinear relativistic second-order differential equation for the electron-fluid in Lagrangian coordinates was derived. This allowed to study the dynamics until wave-breaking in closer detail by numerical integration. By the Lagrangian coordinates description inhomogeneities in the background density were identified to be a possible source of wave-breaking. This result is not based on relativistic effects.

The instabilities of 1D solitons clearly show wave-breaking in a homogeneous plasma as part of the nonlinear evolution, but the excited fields do not always match the Dawson-criterion. Obviously the criteria for wave-breaking have to be refined.

We will make use of the Lagrangian coordinate framework to study the influence of relativistic nonlinearities on the stability of wake-fields. We will extend the breaking-criterion to the relativistic regime and see that breaking occurs without threshold. However, the time-scale on which breaking takes place may be very long. An estimate for the breaking time will be given.

The derivation of the 1D solitons assumes that all quantities are constant transversely to their propagation direction. In simulations and experiments however localized structures are found. We expect to observe a transition from plane 1D solitons into localized pulse filaments by a transversal instability. The effect of transversal instability of solitons is well known from classical solitons [46]. To analyze this scenario we have to allow for a transversal dependence of all quantities. Since this transversal direction is arbitrary, we choose a system where the transversal direction is along a single transversal coordinate. This means we have a two-dimensional (2D) Maxwell-fluid description.

Within this 2D model we will investigate the transversal instability of circular polarized solitons in cold electron-ion plasma. We will employ the same stability analysis technique for 2D geometry as we used in 1D. The rate of instability will depend on the wave-number $k_\perp$ of the transversal perturbation. We will quantify this dependence and find the fastest growing perturbation.

Following the slaving principle [34], the most unstable mode may dominate the nonlinear evolution, slave all others and show up in the topology of the nonlinear end state. To verify if this principle is at work here, we carry out nonlinear 2D simulations. The structure of the 2D instability may already give us a hint to dynamics in higher dimensions.

The field structure of the fastest growing perturbation will be analyzed in terms

1. Introduction

of polarization. The results from this discussion will be compared to results from literature, where the polarization of solitons created from linear polarized lasers is discussed [59].

The fully three-dimensional (3D) study of instabilities is not feasible yet due to limitations in computing power, however we will present that it is possible to gain insight into the weakly perturbed 3D regime from the 2D results.

The organization of this book is the following. In the next chapter the Maxwell-fluid model describing the laser potential and the plasma response is derived. The numerical methods to simulate the model equations are discussed at the end of chapter 2. In chapter 3 a numerical technique for stability analysis is developed. This method will be used to study stability of different relativistic 1D solitons in chapter 4. Chapter 4 will cover stability of pre-solitons and post-solitons in 1D. Wave-breaking due to relativistic effects will be discussed in chapter 5. Chapter 6 covers the influence of transversal perturbations on plane relativistic solitons. The work is summarized by a conclusion in chapter 7.

2 Physical model

To describe the laser and the plasma response, we will use a Maxwell-fluid description. It consists of fluid equations for the electrons and the ions, coupled to the Maxwell equations for the vector potential $\mathbf{A}$ and the scalar potential ϕ. The fluid description of the plasma components does not include kinetic effects such as particle acceleration and plasma heating. The advantage of fluid models over kinetic models on the other hand is the lower computational effort to simulate them and much less numerical noise.

First we will discuss the relativistic Maxwell-fluid equations in 3D geometry, in later chapters we will derive reduced versions for 2D and 1D geometry.

2.1 Maxwell-fluid equations

The Maxwell-equations for the vector potential $\mathbf{A}$ and scalar potential ϕ in Coulomb gauge are

$$\frac{1}{c^2}\frac{\partial^2}{\partial t^2}\mathbf{A} - \Delta\mathbf{A} = \frac{4\pi}{c}\mathbf{j} - \frac{1}{c}\frac{\partial}{\partial t}\nabla\phi, \tag{2.1}$$

$$\Delta\phi = -4\pi\rho. \tag{2.2}$$

The fluid equations are coupled to these equations via the densities and the momenta of the plasma species. For electrons and protons with charge $q_e = e$ and $q_i = -e$ the

2. Physical model

current density $\mathbf{j}$ and charge density ρ are given by

$$\mathbf{j} = e(n_i \mathbf{v}_i - n_e \mathbf{v}_e), \tag{2.3}$$

$$\rho = e(n_e - n_i), \tag{2.4}$$

with velocities $\mathbf{v}_\alpha$ and densities n_α for the species α ($\alpha = e, i$). Let $\mathbf{p}_\alpha$ be the momentum of species α, the density and the momentum balances are given by

$$\frac{\partial}{\partial t} n_\alpha + \nabla \cdot (n_\alpha \mathbf{v}_\alpha) = 0, \tag{2.5}$$

$$\frac{\partial}{\partial t} \mathbf{p}_\alpha + (\mathbf{v}_\alpha \cdot \nabla) \mathbf{p}_\alpha = q_\alpha \left[-\nabla \phi - \frac{1}{c} \frac{\partial}{\partial t} \mathbf{A} + \frac{1}{c} \mathbf{v}_\alpha \times (\nabla \times \mathbf{A}) \right] - \frac{1}{n_\alpha} \nabla \cdot \Pi_\alpha. \tag{2.6}$$

We shall suppose an isotropic plasma, in this case all off-diagonal entries of the pressure tensor Π_α are zero and all diagonal entries are equal. Let P_α be the entry on the diagonal, then $\nabla \cdot \Pi_\alpha = \nabla P_\alpha$. The system (2.1)-(2.6) should be closed by an equation of state for the scalar pressure P_α. We suppose the plasma to behave like an ideal gas, hence $P_\alpha = n_\alpha k_B T_\alpha$, with temperatures T_α and Boltzmann constant k_B. To close the system of equations we consider an isothermal plasma with $T_\alpha = const$.

The quiver velocities $\mathbf{v}_\alpha$ of particles oscillating in an electromagnetic field are related to the kinetic momenta $\mathbf{p}_\alpha$ by

$$\mathbf{v}_a = \frac{\mathbf{p}_\alpha}{m_\alpha \gamma_\alpha}, \tag{2.7}$$

where

$$\gamma_\alpha = \frac{1}{\sqrt{1 - \left(\frac{v_\alpha}{c}\right)^2}} = \sqrt{1 + \left(\frac{\mathbf{p}_\alpha}{m_\alpha c}\right)^2}, \tag{2.8}$$

with m_α the rest mass of species α.

The momentum balance can be rewritten as a balance equation for the canonical

2.1. Maxwell-fluid equations

momentum $\mathbf{M}_\alpha$,

$$\frac{\partial}{\partial t}\mathbf{M}_\alpha = -q_\alpha \nabla\phi - m_\alpha c^2 \nabla \gamma_\alpha + \frac{1}{m_\alpha \gamma_\alpha}\mathbf{p}_\alpha \times (\nabla \times M_\alpha) - \frac{T_\alpha k_B}{n_\alpha}\nabla n_\alpha \qquad (2.9)$$

with $\mathbf{M}_\alpha = \mathbf{p}_\alpha - \frac{q_\alpha}{c}\mathbf{A}$.

Throughout this work, all quantities are normalized. Normalization is done by $x \to x\omega_{pe}/c$, $t \to t\,\omega_{pe}$, $\mathbf{v}_\alpha \to \mathbf{v}_\alpha/c$, $\mathbf{p}_\alpha \to \mathbf{p}_\alpha/(m_e c)$, $\phi \to e\phi/(m_e c^2)$, $\mathbf{A} \to e\mathbf{A}/(m_e c^2)$, $n_\alpha \to n_\alpha/n_0$, $T_\alpha \to T_\alpha/(m_e c^2)$. Here, $\omega_{pe} = (4\pi n_0 e^2/m_e)^{1/2}$ is the electron plasma frequency and n_0 the unperturbed electron (and ion) density; $\omega = 1$ defines the critical density. We normalize the charges by the electron charge e, so $q_e = 1$, and $q_i = -1$. We introduce the mass ratios $\varepsilon_e = 1$, $\varepsilon_i = m_e/m_i$.

The dimensionless Maxwell-fluid equations in Coulomb gauge are given by

$$\frac{\partial^2}{\partial t^2}\mathbf{A} - \Delta\mathbf{A} = \mathbf{j} - \frac{\partial}{\partial t}\nabla\phi, \qquad (2.10)$$

$$\Delta\phi = n_e - n_i, \qquad (2.11)$$

$$\frac{\partial n_\alpha}{\partial t} + \nabla \cdot (n_\alpha \mathbf{v}_\alpha) = 0, \qquad (2.12)$$

$$\frac{\partial}{\partial t}\mathbf{M}_\alpha - \frac{\varepsilon_\alpha}{\gamma_\alpha}\mathbf{p}_\alpha \times (\nabla \times \mathbf{M}_\alpha) = \nabla\left(q_\alpha\phi - \frac{\gamma_\alpha}{\varepsilon_\alpha}\right) - \frac{T_\alpha}{n_\alpha}\nabla n_\alpha, \qquad (2.13)$$

$$\mathbf{j} = \varepsilon_i \frac{n_i \mathbf{p}_i}{\gamma_i} - \frac{n_e \mathbf{p}_e}{\gamma_e}, \qquad (2.14)$$

with

$$\gamma_e = \sqrt{1 + |\mathbf{p}_e|^2}, \qquad (2.15)$$

$$\gamma_i = \sqrt{1 + \varepsilon_i^2 |\mathbf{p}_i|^2}. \qquad (2.16)$$

To reduce the numerical effort to solve equation (2.10) and provide a more robust algorithm, we introduce the two projection operators P^{cf} and P^{df} such that a vector

2. Physical model

field $\mathbf{u}$ is decomposed into $\mathbf{u} = \mathbf{v} + \mathbf{w}$ with the properties

$$\mathsf{P}^{cf}\mathbf{u} = \mathbf{v} \equiv \mathbf{u}^{cf}, \quad \nabla \times \mathbf{v} = 0, \quad \text{but generally } \nabla \cdot \mathbf{v} \text{ not equal zero}, \tag{2.17}$$

$$\mathsf{P}^{df}\mathbf{u} = \mathbf{w} \equiv \mathbf{u}^{df}, \quad \nabla \cdot \mathbf{w} = 0, \quad \text{but generally } \nabla \times \mathbf{w} \text{ not equal zero} \tag{2.18}$$

with

$$\mathsf{P}^{df} + \mathsf{P}^{cf} = \mathbf{1}. \tag{2.19}$$

The operators are represented by

$$\mathsf{P}^{df} = \mathbf{1} - \nabla(\Delta^{-1})\nabla \cdot \quad \text{and} \quad \mathsf{P}^{cf} = \nabla(\Delta^{-1})\nabla \cdot . \tag{2.20}$$

Applying these operators to Eq. (2.10) yields expressions for the curl-free and the divergence-free part

$$\frac{\partial^2}{\partial t^2}\mathbf{A} - \Delta\mathbf{A} = \mathbf{j}^{df}, \tag{2.21}$$

$$\frac{\partial}{\partial t}\nabla\phi = \mathbf{j}^{cf}. \tag{2.22}$$

Writing the kinetic momenta as $\mathbf{p}_\alpha = \mathbf{p}_\alpha^{df} + \mathbf{p}_\alpha^{cf}$ and splitting of the momentum balance into curl-free and divergence-free parts leads to

$$\frac{\partial}{\partial t}\left(\mathbf{p}_\alpha^{df} - \mathbf{A}\right) - \mathsf{P}^{df}\left[\frac{\varepsilon_\alpha}{\gamma_\alpha}\mathbf{p}_\alpha \times \left(\nabla \times \left(\mathbf{p}_\alpha^{cf} - \mathbf{A}\right)\right)\right] = 0, \tag{2.23}$$

$$\frac{\partial}{\partial t}\mathbf{p}_\alpha^{cf} - \mathsf{P}^{cf}\left[\frac{\varepsilon_\alpha}{\gamma_\alpha}\mathbf{p}_\alpha \times \left(\nabla \times \left(\mathbf{p}_\alpha^{cf} - \mathbf{A}\right)\right)\right] = \nabla\left(q_\alpha\phi - \frac{\gamma_\alpha}{\varepsilon_\alpha}\right) - \frac{T_\alpha}{n_\alpha}\nabla n_\alpha. \tag{2.24}$$

Equation (2.24) implies that for initial conditions $\mathbf{p}_\alpha^{cf} = \mathbf{A}$, the canonical momentum $\mathbf{M}_\alpha$ stays curl-free for all times.

To perform the calculations in the upcoming chapters more efficiently, it is usefull

2.2. Conserved quantities

to consider a co-moving frame of reference. Let x be the direction of propagation and V the velocity of the frame. We introduce $\xi = x - Vt$ and $\tau = t$. In this frame the equations are

$$\frac{\partial^2}{\partial \tau^2}\mathbf{A} - 2V\frac{\partial^2}{\partial \xi \partial \tau}\mathbf{A} + V^2\frac{\partial^2}{\partial \xi^2}\mathbf{A} - \Delta \mathbf{A} = \mathsf{P}^{df}\mathbf{j} , \quad (2.25)$$

$$\Delta \phi = n_e - n_i , \quad (2.26)$$

$$\frac{\partial}{\partial \tau}n_\alpha - V\frac{\partial}{\partial \xi}n_\alpha + \nabla \cdot \mathbf{j} = 0 , \quad (2.27)$$

$$\frac{\partial}{\partial \tau}\mathbf{M}_\alpha - V\frac{\partial}{\partial \xi}\mathbf{M}_\alpha - \frac{\varepsilon_\alpha}{\gamma_\alpha}\mathbf{p}_\alpha \times (\nabla \times \mathbf{M}_\alpha) = \nabla\left(q_\alpha \phi - \frac{\gamma_\alpha}{\varepsilon_\alpha}\right) - \frac{T_\alpha}{n_\alpha}\nabla n_\alpha , \quad (2.28)$$

$$\mathbf{j} = \varepsilon_i \frac{n_i \mathbf{p}_i}{\gamma_i} - \frac{n_e \mathbf{p}_e}{\gamma_e} , \quad (2.29)$$

with $\nabla = (\partial_\xi, \partial_y, \partial_z)^T$.

The complexity of the equations has been reduced compared to the original Mawell-fluid equations. Yet they are still a challenge for numerical treatment, because they contain effects such as shock formation and wave-breaking. The nonlinearities can lead to a crossing of the characteristics of the equations which then leads to the appearance of unphysical negative densities within the Maxwell-fluid model.

2.2 Conserved quantities

The system of equations (2.25) - (2.29) has several conserved quantities. These quantities can be used to verify the accuracy of numerical simulations. We consider a given

2. Physical model

volume $\mathcal{V}$, the total energy W in this volume is conserved and given by

$$W = \frac{1}{8\pi} \int \left\{ E^2 + B^2 + 2n_e(\gamma_e - 1) + 2\frac{n_i}{\varepsilon_i}(\gamma_i - 1) \right\} d\mathcal{V} \tag{2.30}$$

$$= \frac{1}{8\pi} \int \left\{ \left[-\nabla\phi - \left(\frac{\partial}{\partial \tau} - V\frac{\partial}{\partial \xi} \right) \mathbf{A} \right]^2 + (\nabla \times \mathbf{A})^2 \right. \tag{2.31}$$

$$\left. + 2n_e(\gamma_e - 1) + 2\frac{n_i}{\varepsilon_i}(\gamma_i - 1) \right\} d\mathcal{V}. \tag{2.32}$$

The momentum contained in the volume $\mathcal{V}$ is proportional to

$$\mathcal{P} = \int \left\{ n_e \mathbf{p}_e + n_i \mathbf{p}_i + \mathbf{E} \times \mathbf{B} \right\} d\mathcal{V}, \tag{2.33}$$

where each component is conserved on its own for $\mathcal{P}$. The total particle number

$$N_\alpha = \int n_\alpha d\mathcal{V} \tag{2.34}$$

has to be conserved for each species $\alpha = e, i$.

2.3 Numerical Algorithms

Let us discuss the numerical methods used in the following chapters to solve the nonlinear and the linearized equations. We will have to solve the equations in different geometries (1D and 2D), which will be introduced in the next chapters. Common to all applications is that the equations are coupled partial differential equations in time and space. All used codes are based on a second-order finite-difference scheme in time. The advancement in time is determined by a Leap-Frog step. Let $u = u(t)$, then

$$\frac{\partial}{\partial t} u(t) \to \frac{u(t + \Delta t) - 2u(t) + u(t - \Delta t)}{(\Delta t)^2}. \tag{2.35}$$

2.3. Numerical Algorithms

For spatial derivatives and integration spectral methods are used. Spatial derivatives of a function $f(x)$ can be calculated by multiplying the Fourier-transformation (FT) $\hat{f}$ of f by $-\mathrm{i}k$ and performing the backward Fourier-transformation, i.e.

$$\frac{\partial}{\partial x} f(x) \to \mathrm{FT}^{-1}\left(-\mathrm{i}k_x \, \mathrm{FT}(f(x))\right) . \tag{2.36}$$

We deal with discrete distributions in our simulations, hence the discretized version of Eq. (2.36) is implemented. It uses the discrete Fourier-transformation (DFT). Compared to finite difference schemes, the spectral scheme produces vastly lower numerical errors when differentiating discrete distributions. The distribution to be differentiated has to fulfill some assumptions like e.g. quasi-periodicy for the discrete Fourier transformation (DFT).

Spatial integration of distributions can be done with spectral methods, too. This is for example required to recover the potential ϕ or the electric field $E = -\nabla \phi$ from the densities n_e and n_i via the Laplace equation $\Delta \phi = n_e - n_i$. In order to integrate a function $f(x)$, we perform a discrete Fourier-transformation of the distribution, multiply every mode by i/k_x and perform the inverse discrete Fourier-transformation, i.e.

$$\int f(x)\,\mathrm{d}x \to \mathrm{DFT}^{-1}\left(\frac{\mathrm{i}}{k_x} \mathrm{DFT}(f(x))\right) . \tag{2.37}$$

The treatment of the $k_x = 0$ mode requires additional assumptions about boundary conditions.

For the 2D simulations we need to implement the operator $\mathsf{P}^{df} = 1 - \nabla(\Delta^{-1})\nabla \cdot$ which gives the divergence-free part of a vector field. We choose the two spatial coordinates to be x and y. Let $\hat{u}_j$ be the DFT of the j component of the vector field

2. Physical model

u. The j component of divergence-free part of the vector field **u** is given by

$$(\mathsf{P}^{df}\mathbf{u})_j = \mathrm{DFT}^{-1}\left(\hat{u}_j - \mathrm{i}k_j \frac{1}{-k_x^2 - k_y^2}(-\mathrm{i}k_x \hat{u}_x - \mathrm{i}k_y \hat{u}_y)\right). \tag{2.38}$$

Again the $k = 0$ mode has to be treated by making assumptions about the resulting distributions. In our case we assume the amplitude of this mode to be 0 in order to meet the boundary conditions.

The discrete Fourier transformation used in the implementation is provided by the Intel MKL library. Simulations in 2D benefit by its multithreaded implementation of the DFT, which makes use of multiple CPU cores to execute a 2D transformation.

When using spectral methods we have to anticipate aliasing effects [30]. To keep aliasing errors as small as possible and still retain a large as possible correctly treated maximum wave number we apply a filter function in Fourier space that cuts 10 % of the highest modes. This cutting is only done for the densities n_e and n_i, which proved to be sufficient to avoid numerical instabilities due to aliasing.

In literature codes are described that use finite difference schemes in space, too. These codes are much more limited in their numerical stability. The numerical errors due to finite differences in space may serve as initial conditions for a numerical or physical instability [75].

For all nonlinear simulations, it is possible to check their stability by the conservation of the conserved quantities listed in Sec. 2.2. All presented nonlinear simulations have been checked for the conservation of these quantities to a sensible accuracy.

For the linearized Maxwell-fluid equations (3.14)-(2.29) the quantities from Sec. 2.2 do not have to be stationary. In this case we have to find solutions which are stationary in order to check the correctness of the code. Let us suppose we know a stationary solution for the system (2.25)-(2.29) which is given by $\mathbf{A}_0$, $n_{\alpha 0}$ and $\mathbf{p}_{\alpha 0}$. A stationary solution to the linearized Maxwell-fluid equations (3.14)-(2.29) is given by $\partial \mathbf{A}_0 / \partial \xi$,

2.3. Numerical Algorithms

$\partial n_{\alpha 0}/\partial \xi$ and $\partial \mathbf{p}_{\alpha 0}/\partial \xi$. By using the derivatives as initial conditions, it is possible to check the correctness of the simulation of the linearized Maxwell-fluid equations.

2. Physical model

3 Techniques in stability analysis

3.1 Idea of stability analysis

To exemplify the basic idea of stability analysis let us treat a simple system. We assume a perturbation **p**, which will evolve in time as

$$\frac{\mathrm{d}}{\mathrm{d}t}\mathbf{p} = \mathsf{M}\,\mathbf{p} \tag{3.1}$$

with constant coefficients M_{jk}. Let us assume that the matrix M is well behaved and that the system of eigenvectors is complete. Let $\mathbf{e}_i$ be the eigenvectors and λ_i the eigenvalues of M, i.e.

$$\mathsf{M}\,\mathbf{e}_i = \lambda_i\,\mathbf{e}_i\,. \tag{3.2}$$

Then

$$\phi = \alpha\,\mathbf{e}_i\,e^{\lambda_i t}\,, \quad \alpha = const. \tag{3.3}$$

is a special solution of Eq. (3.1).

As the eigenfunction system of M is complete, we may express any perturbation as

$$\mathbf{p} = \sum_i \alpha_i \mathbf{e}_i\,. \tag{3.4}$$

The eigenvalues λ_i will determine the evolution of each eigenvector. In general they

3. Techniques in stability analysis

are complex, i.e. $\lambda_i = \Gamma + i\omega$. If all λ_i have negative real parts, **p** will vanish after some time, therefore the system which is perturbed is *stable*. If any one of the λ_i has a positive real part, the perturbation will grow without bounds in time, hence the system is *unstable*. In case of all λ_i having a zero real part, but a non-zero imaginary part we will call the system *marginally stable*.

Physically it is in general not necessary to know all eigenvalues and eigenvectors of **M**. It is sufficient to know the eigenvector for the eigenvalue with the largest real part Γ. If this Γ is positive, the system is unstable for perturbations in the direction of this eigenvector.

For low-dimensional problems it is possible to analytically derive the eigenvalues and the eigenvectors. For many physical problems the analytical treatment is not feasible due to the size of the matrix **M**. In this case numerical methods have to be applied. There are many possible numerical methods to determine the eigenvalues and eigenvectors; many of these rely on assumptions about the eigenvalues or the eigenvectors.

One possible and very general way to determine the most unstable mode is to solve the system (3.1). Assuming initial conditions

$$\mathbf{p}(t=0) = \sum_i \alpha_i \mathbf{e}_i \tag{3.5}$$

we will get at time t the solution

$$\mathbf{p}(t) = \sum_i \alpha_i e^{\lambda_i t} \mathbf{e}_i . \tag{3.6}$$

This means that after simulating the system for a sufficiently long time, the most unstable mode will eventually dominate all others and we are left with just this mode. The time evolution of this mode determines the growth rate Γ.

3.1. Idea of stability analysis

For time-dependent coefficients the treatment of the linear system is in general much more complicated. Time-dependence of these coefficients may be introduced by a possible time-dependence of the unperturbed state. We are only interested in two special cases here. The first one is where the time-dependence of the coefficients can be transformed out analytically. In this case we can employ the method outlined above for the resulting system. The second one is where there is a periodic dependency of the coefficients. In this case we may use the same method as described above, but in order to calculate the growth rate, we only may measure the magnitude of the perturbation at fixed times that are separated by the period T of the unperturbed state [62].

For nonlinear problems the linear system describing the evolution of a small perturbation is given by linearizing the nonlinear problem about the unperturbed solution.

In case that the system that governs the evolution of the perturbation is not a system of ordinary differential equations (ODEs) like (3.1), but a system of partial differential equations (PDEs) in time and space, we have to transfer it into a system of ODEs. This is done by discretizing all quantities in space.

3.1.1 Numerical linear stability analysis

The system that describes the linear evolution of the perturbation is solved by numerical methods described in Sec. 2.3. If there is a part along the most unstable mode contained within the initial conditions, the amplitude of the distribution will grow in time. As the system is linear, we can divide the distribution by any factor we like at times $t_n = t_n + nT$. In general we reduce the magnitude of the distribution to the order of 1 at these times. This is advantageous for numerics because the accumulated numerical errors will be reduced in size by this method, too. If the coefficients for the linear system are periodic with time time T_p, due to a periodic behavior of the unperturbed solution, we choose $T = T_p$.

3. Techniques in stability analysis

3.2 Problems of linear stability analysis

Linear stability analysis implicitly assumes that the perturbation and its spatial derivatives are small compared to the unperturbed solution at every point in the spatial domain.

Not all growing modes found by linear stability analysis are instabilities. An example for this are solitons of the cubic Schrödinger equation

$$i\partial_t \psi + \partial_{xx}\psi + |\psi|^2 \psi = 0, \qquad (3.7)$$

for which the solitons with velocity V are given by

$$\psi(x,t;\eta,V) = \sqrt{2}\,\eta \operatorname{sech}(\eta(x-Vt))\, e^{i\frac{Vx}{2}+i(\eta^2-\frac{V^2}{2})t}. \qquad (3.8)$$

Introducing a perturbation ψ_1 and linearizing Eq. (3.7) about the unperturbed solution ψ_0 we get

$$i\,\partial_t \psi_1 + \partial_{xx}\psi_1 + 2|\psi_0|^2 \psi_1 + \psi_0^2 \psi_1^* = 0. \qquad (3.9)$$

Let ϕ_P be a perturbation of the form

$$\phi_P = \psi(x,t;\eta+\delta\eta,V) - \psi(x,t;\eta,V). \qquad (3.10)$$

Expanding ϕ_P in powers of $\delta\eta$ and stopping with the linear term, we get the so called *phasor-mode*, which results in

$$\phi_P = \operatorname{sech}(\eta(x-Vt))\left(1+\eta^2\left\{2\mathrm{i}\,t - \tanh(\eta(x-Vt))\right\}\right) e^{i\frac{Vx}{2}+i(\eta^2-\frac{V^2}{2})t}. \qquad (3.11)$$

The phasor-mode is a solution of Eq. (3.9) and grows proportional to t. The pertur-

3.2. Problems of linear stability analysis

bation of the second parameter V

$$\phi_T = \psi(x,t;\eta,V+\delta V) - \psi(x,t;\eta,V) \tag{3.12}$$

is the so called *translation-mode*, which is a solution of Eq. (3.9) and growing with t, too. It is given by

$$\phi_T = \text{sech}\left(\eta(x-Vt)\right)\left(\eta\, t\, \tanh(\eta(x-Vt)) + \mathrm{i}\left(\frac{V}{2}-Vt\right)\right) e^{\mathrm{i}\frac{Vx}{2}+\mathrm{i}(\eta^2-\frac{V^2}{2})t}. \tag{3.13}$$

The phasor-mode and the translation-mode are solutions of the linearized Eq. (3.9) for an infinitesimal perturbation in phase or speed. They grow without bounds in time proportional to t. Even though these modes are growing, their growth is not to be understood as an instability. Over time these modes only separate in terms of phase or position from the unperturbed state. It is the method of linearization that implicitly carries an unphysical concept of stability. The more physical concept which is appropriate in this place is the stability of invariant sets, see Appendix A.1.

In the following chapters we are going to investigate for stability of solitons which depend on parameters like frequency, velocity or temperature. The parameter dependence of the solitons is a possible source for growing modes that are analogous to the phasor- and the translation-mode in the example above. In most cases such modes grow according to a power-law and not exponentially. In order to be sure that the most unstable mode we get from the linear system is an instability, we check our findings by nonlinear simulations. The initial conditions used for these simulations are the unperturbed state plus a small amount of the mode we found in the linear case. As long as the perturbation is only a small part of the whole numerical solution, it has to grow according to $e^{\Gamma t}$ to be an instability.

3. Techniques in stability analysis

3.3 Linearized Maxwell-fluid equations

To follow the linear evolution of a perturbation of relativistic solitons we have to linearize the Maxwell-fluid equations (2.25) - (2.29) about an unperturbed solution. Let the unperturbed solution consist of $\mathbf{A}_0$, $\mathbf{p}_{\alpha 0}$, $n_{\alpha 0}$. We introduce perturbations $\mathbf{A}_1$, $\mathbf{p}_{\alpha 1}$, $n_{\alpha 1}$ such that $\mathbf{A} = \mathbf{A}_0 + \epsilon \mathbf{A}_1$, $\mathbf{p}_\alpha = \mathbf{p}_{\alpha 0} + \epsilon \mathbf{p}_{\alpha 1}$, $n_\alpha = n_{\alpha 0} + \epsilon n_{\alpha 1}$, where ϵ is a smallness parameter.

We will perform linear stability calculations only in cold plasmas $T = 0$. Subsequent linearization of the model equations leads to

$$\frac{\partial^2}{\partial \tau^2}\mathbf{A}_1 - 2V\frac{\partial^2}{\partial \xi \partial \tau}\mathbf{A}_1 + V^2\frac{\partial^2}{\partial \xi^2}\mathbf{A}_1 - \Delta \mathbf{A}_1 = \mathsf{P}^{df}\left(\mathbf{j}_{i1} - \mathbf{j}_{e1}\right) , \quad (3.14)$$

$$\Delta \phi_1 = n_{e1} - n_{i1} , \quad (3.15)$$

$$\frac{\partial}{\partial \tau}n_{\alpha 1} - V\frac{\partial}{\partial \xi}n_{\alpha 1} + \nabla \cdot \mathbf{j}_{\alpha 1} = 0 , \quad (3.16)$$

$$\mathbf{j}_{\alpha 1} = \frac{\epsilon_\alpha}{\gamma_{\alpha 0}}\left[n_{\alpha 1}\mathbf{p}_{\alpha 0} + n_{\alpha 0}\mathbf{p}_{\alpha 1} - \frac{1}{\gamma_{\alpha 0}}n_{\alpha 0}\gamma_{\alpha 1}\mathbf{p}_{\alpha 0}\right] , \quad (3.17)$$

$$\frac{\partial}{\partial t}\mathbf{M}_{\alpha 1} - V\frac{\partial}{\partial x}\mathbf{M}_{\alpha 1}$$
$$- \frac{\epsilon_\alpha}{\gamma_{\alpha 0}}\left[-\frac{1}{\gamma_{\alpha 0}}\gamma_{\alpha 1}\mathbf{p}_{\alpha 0} \times (\nabla \times \mathbf{M}_{\alpha 0}) + \mathbf{p}_{\alpha 1} \times (\nabla \times \mathbf{M}_{\alpha 0}) + \mathbf{p}_{\alpha 0} \times (\nabla \times \mathbf{M}_{\alpha 1})\right]$$
$$= \nabla\left(q_\alpha \phi_1 - \frac{1}{\epsilon_\alpha}\gamma_{\alpha 1}\right) , \quad (3.18)$$

with

$$\gamma_{\alpha 1} = \epsilon_\alpha^2 \frac{\mathbf{p}_{\alpha 0} \cdot \mathbf{p}_{\alpha 1}}{\gamma_{\alpha 0}} . \quad (3.19)$$

3.3. Linearized Maxwell-fluid equations

The set of equations (3.14)-(3.18) is the basis for the numerical stability analysis of different soliton solutions in the upcoming chapters. It will be applied in different geometries and on different time-scales (e.g. neglecting variation of ion density and assuming zero parallel ion momentum for pre-solitons).

3. Techniques in stability analysis

4 Stability and dynamics of relativistic 1D solitons

In the introduction we discussed the physical processes that lead to the formation of solitary structures during relativistic laser-plasma interaction. Besides simulations that show the creation process of solitary structures and the later evolution, a number of analytical approaches has been taken to get to an analytical understanding of relativistic solitons [42, 26, 25, 66, 32, 57, 72, 73, 58].

In this chapter, we will discuss the *longitudinal* stability of known stationary solutions of the Maxwell-fluid equations. This will include circular polarized soliton solutions on the electron- and on the ion-timescale [42, 25, 23], linear polarized solitons on the electron-timescale [32] and circular polarized solitons in warm electron-ion-plasma [58]. Our numerical stability analysis will provide us the structure of the fastest growing mode and its growth rate Γ. The growth rate will determine the physical importance of the instability. Fully nonlinear simulations allow us to make prediction for the nonlinear end state that we would expect.

Let us assume a laser pulse propagating in x direction. If the focal spot size, i.e. the transversal dimension of the laser pulse, is much larger than the length of the pulse in propagation direction, we can assume to a good approximation that all quantities depend just on the coordinate x in propagation direction and time, but not on the transversal coordinates y and z.

4. Stability and dynamics of relativistic 1D solitons

4.1 1D model equations

The assumption of transversely constant quantities reduces the general 3D model (2.25) - (2.29) to the standard 1D model. We arrive at this model by neglecting derivatives along the transversal directions y and z, so that $\nabla = (\partial_\xi, 0, 0)^T$ in (2.25) - (2.29). The Coulomb gauge $\nabla \cdot \mathbf{A} = 0$ now requires $A_x = 0$. The vector potential is therefore purely transversal

$$\mathbf{A} = (0, A_y, A_z)^T. \tag{4.1}$$

We introduce a complex notation of the transversal components

$$A_\perp = A_y + \mathrm{i} A_z. \tag{4.2}$$

The momenta are split into a parallel and a transversal part $\mathbf{p}_\alpha = p_\alpha \mathbf{e}_x + \mathbf{p}_{\alpha\perp}$. For the derivation of all 1D soliton solutions $\mathbf{p}_{\alpha\perp} = q_\alpha \mathbf{A}_\perp$ is supposed, thus $\mathbf{M}_{\alpha\perp} = 0$.

All together the relativistic model equations in 1D geometry are

$$\frac{\partial^2}{\partial \tau^2} A_\perp - 2V \frac{\partial^2}{\partial \xi \partial \tau} A_\perp - \left(1 - V^2\right) \frac{\partial^2}{\partial \xi^2} A_\perp = -\left(\varepsilon_i \frac{n_i}{\gamma_i} + \frac{n_e}{\gamma_e}\right) A_\perp, \tag{4.3}$$

$$\frac{\partial^2}{\partial \xi^2} \phi = n_e - n_i, \tag{4.4}$$

$$\frac{\partial}{\partial \tau} n_\alpha - V \frac{\partial}{\partial \xi} n_\alpha + \frac{\partial}{\partial \xi} j_{\alpha x} = 0, \tag{4.5}$$

$$\frac{\partial}{\partial \tau} p_\alpha - V \frac{\partial}{\partial \xi} p_\alpha = \frac{\partial}{\partial \xi} \left(q_\alpha \phi - \frac{\gamma_\alpha}{\varepsilon_\alpha}\right) - \frac{T_\alpha}{n_\alpha} \frac{\partial}{\partial \xi} n_\alpha, \tag{4.6}$$

4.1. 1D model equations

where

$$\gamma_\alpha = \sqrt{1 + \varepsilon_\alpha^2 \left(|A_\perp|^2 + p_\alpha^2\right)}, \qquad (4.7)$$

$$j_{\alpha\xi} = \varepsilon_\alpha n_\alpha \frac{p_\alpha}{\gamma_\alpha}. \qquad (4.8)$$

Besides to discuss the interplay between backward and forward Raman scattering, modulation of broad light pulses, down-cascade in the frequency spectrum, photon condensation, and break-up of the original laser beam [66], this model is also suitable to investigate 1D slow solitons on the electron time-scale.

The linearized equations (same notation as in Sec.3.3) in 1D are

$$\frac{\partial^2}{\partial \tau^2} A_{\perp 1} - 2V \frac{\partial^2}{\partial \xi \partial \tau} A_{\perp 1} - \left(1 - V^2\right) \frac{\partial^2}{\partial \xi^2} A_{\perp 1} = \\ \sum_\alpha q_\alpha \left\{ \frac{\varepsilon_\alpha}{\gamma_{\alpha 0}} \left(n_{\alpha 1} - \frac{\gamma_{\alpha 1}}{\gamma_{\alpha 0}} n_{\alpha 0} \right) A_{\perp 0} + n_{\alpha 0} A_{\perp 1} \right\}, \qquad (4.9)$$

$$\frac{\partial^2}{\partial \xi^2} \phi_1 = n_{e1} - n_{i1}, \qquad (4.10)$$

$$\frac{\partial}{\partial \tau} n_{\alpha 1} - V \frac{\partial}{\partial \xi} n_{\alpha 1} = -\frac{\partial}{\partial \xi} \varepsilon_\alpha \left(n_{\alpha 1} \frac{p_{\alpha 0}}{\gamma_{\alpha 0}} + p_{\alpha 1} \frac{n_{\alpha 0}}{\gamma_{\alpha 0}} + \frac{\varepsilon_\alpha^2}{\gamma_{\alpha 0}} \mathbf{P}_{\alpha 0} \cdot \mathbf{P}_{\alpha 1} \right), \qquad (4.11)$$

$$\frac{\partial}{\partial \tau} p_{\alpha 1} - V \frac{\partial}{\partial \xi} p_{\alpha 1} = \frac{\partial}{\partial \xi} \left(q_\alpha \phi_1 - \varepsilon_\alpha \frac{\mathbf{P}_{\alpha 0} \cdot \mathbf{P}_{\alpha 1}}{\gamma_{\alpha 0}} \right). \qquad (4.12)$$

For linear polarization we derive a reduced description of the system (4.3) - (4.6), valid for only a weak electron density response and weakly relativistic laser pulse amplitudes. We will treat only standing solitons in this case, hence $V = 0$. Variations of n_i and $p_{i\parallel}$ are neglected, hence the model is only valid on the electron time-scale. We introduce the electron density variation δn_e as $n_e = 1 + \delta n_e$ and suppose $|A_\perp| \ll 1$.

4. Stability and dynamics of relativistic 1D solitons

The wave equation (4.3) reduces to

$$\frac{\partial^2}{\partial \xi^2} A_\perp - \frac{\partial^2}{\partial \tau^2} A_\perp = (1 + \delta n_e) \frac{A_\perp}{\gamma_e} . \tag{4.13}$$

The plasma response is determined by the equations

$$\frac{\partial^2}{\partial \xi^2} \phi = \delta n_e , \tag{4.14}$$

$$\frac{\partial}{\partial \tau} \delta n_e = - \frac{\partial}{\partial \xi} \frac{(1 + \delta n_e) p_{e\|}}{\gamma_e} , \tag{4.15}$$

$$\frac{\partial}{\partial \tau} p_{e\|} = \frac{\partial}{\partial \xi} (\phi - \gamma_e) . \tag{4.16}$$

Let ϵ characterize the amplitude of $A_\perp$. Equation (4.16) suggests the ordering $\delta n_e \sim p_{e\|} \sim a^2 \sim \mathcal{O}(\epsilon^2)$. We take the time derivative of (4.15) and insert expressions (4.14) and (4.16). We keep only terms of the order of ϵ^2, all higher terms are neglected. The result is

$$\frac{\partial^2}{\partial \tau^2} \delta n_e + \delta n_e = \frac{\partial^2}{\partial \xi^2} \gamma_e , \tag{4.17}$$

where $\gamma_e \approx \sqrt{1 + |A_\perp|^2}$ (since $p_{e\|} \ll A_\perp$).

A further simplification of the equations (4.13) and (4.17) is possible by expanding γ_e appropriately. To lowest order we get from (4.13) and (4.17)

$$\left(\frac{\partial^2}{\partial \xi^2} - \frac{\partial^2}{\partial \tau^2} \right) A_\perp = - \left(1 + \delta n_e - \frac{A_\perp^2}{2} + \frac{1}{8} A_\perp^4 \right) A_\perp , \tag{4.18}$$

$$\frac{\partial^2}{\partial \tau^2} \delta n_e + \delta n_e = \frac{1}{2} \frac{\partial^2}{\partial \xi^2} A_\perp^2 . \tag{4.19}$$

The latter model has been used in [32] to derive linear polarized soliton solutions, which we are going to study in Sec. 4.2.2 of this chapter.

4.2 Solitons on the electron time scale

4.2.1 Circular polarization

One of the first approaches to derive soliton solutions in 1D geometry was by Esirkepov et al. [23]. To simplify the equations a static ion background was supposed. Starting from (4.3)-(4.6) and neglecting the ion density variation, standing (V=0) circular polarized solutions where found in the form of ordinary partial differential equations for the vector potential and the momentum. For standing solitons an analytic expression was derived

$$A_\perp = \frac{2\sqrt{1-\omega_0^2}\cosh(\xi\sqrt{1-\omega_0^2})}{\cosh^2(\xi\sqrt{1-\omega_0^2}) + \omega_0^2 - 1} e^{i\omega_0\tau}. \qquad (4.20)$$

The maximum amplitude A_0 and the frequency ω_0 are related by

$$A_0 = \frac{2\sqrt{1-\omega_0^2}}{\omega_0^2}. \qquad (4.21)$$

The frequency ω_0 has to be lower than the electron plasma frequency to keep the radiation trapped, i.e. $\omega_0 < 1$. The maximum amplitude these solitons can achieve is $A_0 = \sqrt{3}$, Fig. 4.1 displays this solution.

By simulation of the linearized equations (4.9)-(4.12) we found that these solitons are stable with respect to small initial perturbations.

4.2.2 Linear polarization

For linear polarization it is much more difficult to solve the fully relativistic 1D model due to the generation of higher harmonics of the incident wave. A simplified approach is to solve, instead of the fully relativistic system (4.3)-(4.6), the weakly nonlinear

4. Stability and dynamics of relativistic 1D solitons

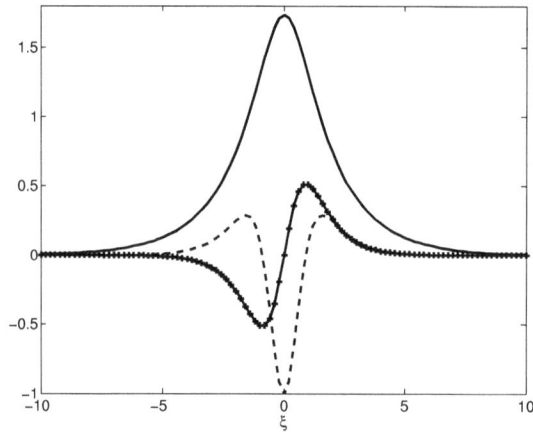

Figure 4.1: Circular polarized soliton with $V = 0$ described by (4.20) with $\omega = \sqrt{2/3}$. The solid line corresponds to the intensity $|A_\perp|^2$, the crossed line to the longitudinal electric field and the dashed line to the density variation $n - n_0$ by the soliton, respectively.

4.2. Solitons on the electron time scale

system (4.18)-(4.19). Introducing a slowly-varying complex envelope in the form

$$A_\perp = \frac{1}{2}\left(\alpha e^{-iE\tau} + \alpha^* e^{iE\tau}\right), \tag{4.22}$$

$$\delta n_e = N_0 + \frac{1}{2}\left(N_2 e^{-2iE\tau} + N_2^* e^{2iE\tau}\right), \tag{4.23}$$

and substituting (4.22) and (4.23) into (4.19) leads to

$$N_0 = \frac{1}{4}|\alpha_\xi|^2, \quad N_2 = -\frac{1}{12}(\alpha^2)_{\xi\xi}. \tag{4.24}$$

With $\eta^2 = 1 - E^2$ we get an equation for $\alpha(\xi)$ in the form

$$(\alpha_\xi)^2 = \frac{\eta^2 \alpha^2 \left(1 - \frac{3}{16}\frac{\alpha^2}{\eta^2} + \frac{5}{192}\frac{\alpha^4}{\eta^2}\right)}{1 - \frac{5}{12}\alpha^2}. \tag{4.25}$$

This equation can be solved numerically and leads to solutions for $A_\perp$ and δn_e.

Depending on the η-value, the solutions are stable or not (on the electron time-scale). Previous investigations [32, 63] predicted stability over a broad range of η-values. The predictions were made by using the Vakitov-Kolokolov criterion [88], see Appendix B.1. The Vakitov-Kolokolov criterion states that solutions are stable with respect to small perturbations as long as $dP/d\eta^2 > 0$, where P is given by

$$P = \int |\alpha(\xi)|^2 d\xi. \tag{4.26}$$

Our numerical simulations of the system (4.18)-(4.19) show that the range of η in which stability prevails is much smaller than as previously predicted on basis of the Vakitov-Kolokolov criterion [32].

Let $A_{\perp 0}$ be the maximum amplitude of the linear polarized soliton. Previous investigations [32] predicted instability only for high intensities $A_{\perp 0} \geq 1.44$. Here, we

4. Stability and dynamics of relativistic 1D solitons

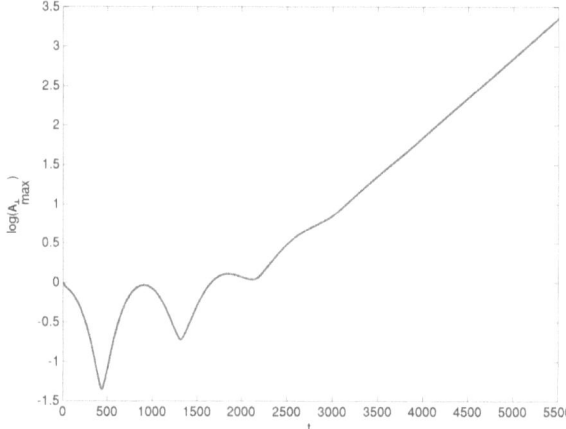

Figure 4.2: Time evolution of the maximum of $A_\perp$ due to numerical integration of the linearized equations, for a linear polarized soliton with $A_{\perp 0} = 0.2$.

4.2. Solitons on the electron time scale

find that the instability sets in already for $A_{\perp 0} \approx 0.2$, i.e. well below the ultra-relativistic regime. The just mentioned threshold amplitude follows within the model (4.18)-(4.19). It is also found in the simulations of the fully relativistic Eqs. (4.3)-(4.6).

The "numerical proof of instability" for amplitudes of the order of, or larger than, $A_{\perp 0} \approx 0.2$ will now be demonstrated on the following graphs. Figure 4.2 shows the time evolution of the maximum of $A_\perp$ due to numerical integration of the linearized versions of equations (4.18)-(4.19), for a linear polarized soliton with $A_{\perp 0} = 0.2$. The most unstable mode can be identified. Figure 4.3 shows the corresponding fastest growing mode. The exponential growth rate Γ for this mode is approximately 0.001. When a significant part of this mode is put as an initial perturbation into the nonlinear integrator, the instability can be clearly seen. Figure 4.4 shows results from the nonlinear integration of an unperturbed and a perturbed linear polarized initial condition, respectively, for $A_{\perp 0} = 0.2$ after the time $t = 1500$. During the time of integration the unperturbed distribution is practically unchanged while a perturbation of one percent leads to significant unstable behavior.

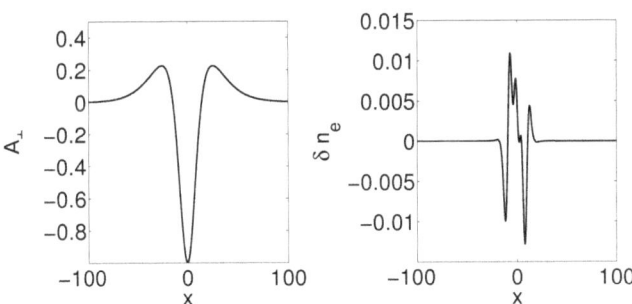

Figure 4.3: Fastest growing mode of the perturbation for a linear polarized soliton with $A_{\perp 0} = 0.2$. The exponential growth rate Γ for this mode is 0.001.

4. Stability and dynamics of relativistic 1D solitons

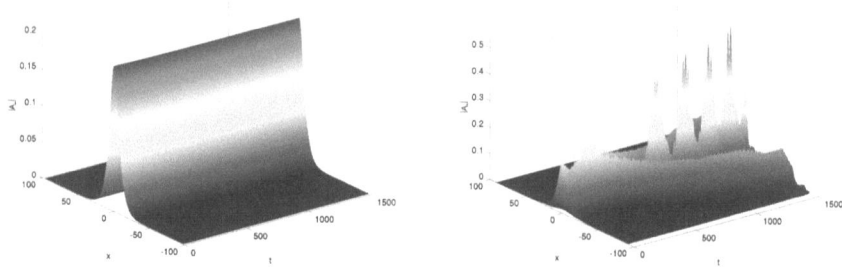

Figure 4.4: Left: Evolution of an unperturbed linear polarized soliton with $A_{\perp 0} = 0.2$. Right: Evolution of a perturbed linear polarized soliton with $A_{\perp 0} = 0.2$. The initial perturbation was 1% of the fastest growing mode.

4.3 Solitons on the ion time-scale

4.3.1 From pre-soliton to post-soliton

The dimensionless nonrelativistic dispersion relation for an electron plasma wave with frequency ω_0 and wave-number k is given by

$$\omega_0 = \sqrt{1 + k^2}. \qquad (4.27)$$

A laser pulse with frequency $\omega_0 > \omega_{pe}$ is moving with the group velocity in the underdense plasma which is given by

$$v_{gr}(k, \omega_0) = \frac{\partial \omega_0}{\partial k} = \frac{k}{\omega_0} = \sqrt{1 - \frac{1}{\omega_0^2}}. \qquad (4.28)$$

4.3. Solitons on the ion time-scale

The time T it takes the pulse to cover the distance of one wavelength $\lambda = 2\pi/k$ is

$$T = \frac{\lambda}{v_{gr}} = \frac{2\pi}{1 - \frac{1}{\omega_0^2}}. \tag{4.29}$$

For a laser frequency of $\omega_0 = 2$ (this corresponds to $n/n_{cr} = 1/4$) a laser pulse would roughly cover one laser wavelength within a time of 12 in normalized units. The typical time-scales on which the species are able to respond to external forces are given by the inverse of their plasma frequency. Hence the time-scale for ions is about 40 times longer than the time-scale for the electrons (if we a assume e.g. an ionized hydrogen plasma). So for a pulse with just a few laser cycles or even sub-cycle pulses, we can assume the ions to be a static background. For short pulses with frequency ω_0 close to ω_{pe} which are moving at lower speed or are even standing, ions can not be considered static anymore.

To demonstrate the transition from a pre-soliton to a post-soliton, we performed simulations of pre-soliton solutions in a plasma with mobile ions, that are initially uniformly distributed. We demonstrate the transition for a standing circular polarized pre-soliton given by Eq. (4.20) with initial amplitude $A_0 = 0.2$. Figure 4.5 shows the evolution of the electron and ion density. After a time of approximately $1/\omega_{pi}$ the ions react to the field of the soliton and move towards the electrons. The system intents to evolve into a quasi-neutral state. As the ions become pushed out of the center of the soliton, the electrons become even more evacuated. This leads to an almost complete evacuation of density in the middle. At the same time the trapped vector potential gets narrower and steeper. With larger initial amplitudes A_0 the whole process becomes faster.

The electrons, being pushed outwards, pile up at the border of the depression. Very large density gradients are achieved. Then, of course, finite pressure (temperature) effects may come into play. We have shown that finite electron temperatures will

4. Stability and dynamics of relativistic 1D solitons

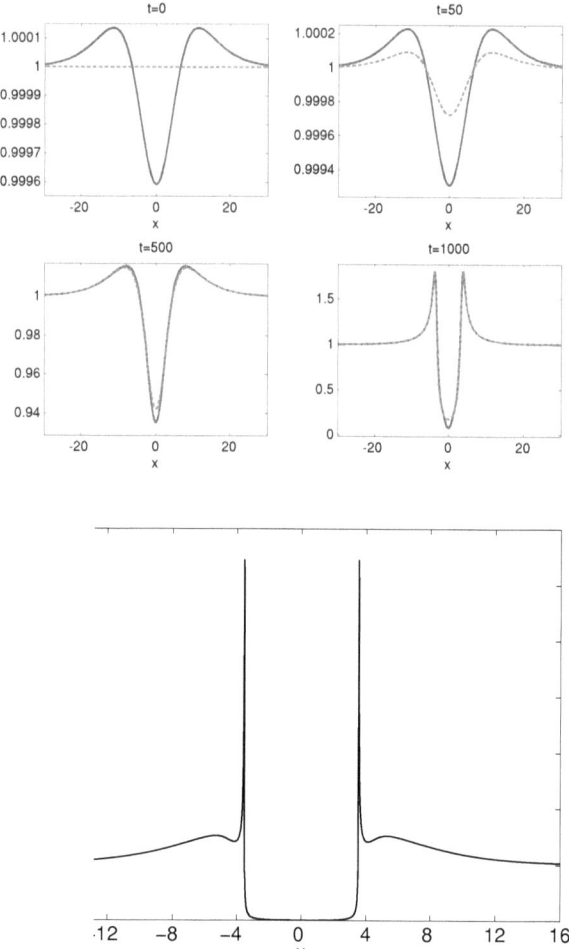

Figure 4.5: Top: Density of ions (dashed line) and electrons (solid line) at different times $t = 0$, $t = 50$, $t = 500$, and $t = 1000$, respectively, when starting with an initial circular polarized pre-soliton with $A_0 = 0.2$ given by Eq. 4.20.
Bottom: Electron density at $t = 1100$ from the same simulation as top figures. Post-soliton created from pre-soltion.

4.3. Solitons on the ion time-scale

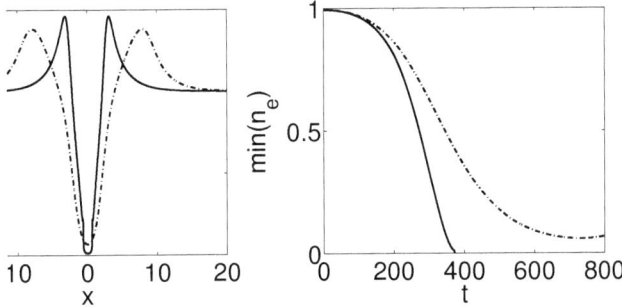

Figure 4.6: Left: Electron density distribution without (solid line) and with (dash-dotted line) electron temperature T_e= 50 keV at times $t = 370$ and $t = 800$, respectively. Right: Evolution of the minimum density in time.

ultimately stop the density steepening. With finite electron temperature no more a complete density evacuation takes place. The required temperatures are quite large, e.g. of order 50 keV. Figure 4.6 shows a typical example. The transition from a pre-soliton to a post-soliton on the ion time-scale was observed experimentally and has already been shown by multi-dimensional PIC simulations [69].

4.3.2 Stationary solitons on the ion time-scale

In general, we have to include ion movement into our system to describe slow solitons. Stationary soliton solutions of (4.3)-(4.6) have been found numerically in Ref. [26] for circular polarization on the basis of the formulation in Ref. [42].

The idea is to express all plasma quantities as functions of the vector potential **A** and the scalar potential ϕ. A system of two coupled ordinary differential equations for the two potentials results, which has to be solved numerically.

4. Stability and dynamics of relativistic 1D solitons

We start by deriving expressions for p_α, γ_α and n_α which only depend on ϕ and $\mathbf{A}$. The vector potential is purely transversal, hence we will combine its two non-vanishing components into one complex quantity $A_\perp = A_y + iA_z$. We introduce $\varrho_\alpha = -q_e m_e/(q_\alpha m_\alpha)$. The boundary conditions are $A_\perp|_{\pm\infty} = 0$, $\phi|_{\pm\infty} = 0$, $p_\alpha|_{\pm\infty} = 0$ and $n_\alpha|_{\pm\infty} = 1$. We integrate (4.5) and (4.6) and assume stationary states, hence $\frac{\partial}{\partial \tau} \to 0$, and get

$$-V \int_{-\infty}^{x} \frac{\partial}{\partial \xi} n_\alpha \mathrm{d}\xi + \int_{-\infty}^{x} \frac{\partial}{\partial \xi} n_\alpha v_\alpha \mathrm{d}\xi = 0, \tag{4.30}$$

$$-V \int_{-\infty}^{x} \frac{\partial}{\partial \xi} p_\alpha \mathrm{d}\xi = -\int_{-\infty}^{x} \frac{\partial}{\partial \xi} q_\alpha \phi \mathrm{d}\xi - \int_{-\infty}^{x} \frac{\partial}{\partial \xi} \frac{\gamma_\alpha}{\varepsilon_\alpha} \mathrm{d}\xi, \tag{4.31}$$

which leads to

$$n_\alpha (v_\alpha - V) = -V, \qquad \text{note } v_\alpha = \varepsilon_\alpha \frac{p_\alpha}{\gamma_\alpha}, \tag{4.32}$$

$$-V p_\alpha = -q_\alpha \phi - \frac{\gamma_\alpha}{\varepsilon_\alpha} + 1. \tag{4.33}$$

The relativistic γ_α factors can be expressed in two ways

$$\gamma_\alpha^2 = 1 + \varepsilon_\alpha^2 \left(A_\perp^2 + p_\alpha^2 \right), \tag{4.34}$$

$$\gamma_\alpha = \varepsilon_\alpha \left(V p_\alpha + 1 - q_\alpha \phi \right). \tag{4.35}$$

We introduce ψ_α and R_α as

$$\psi_\alpha = \varepsilon_\alpha (1 - q_\alpha \phi), \tag{4.36}$$

$$R_\alpha = \sqrt{\psi_\alpha^2 - (1 - V^2)(1 + \varepsilon_\alpha^2 A_\perp^2)}. \tag{4.37}$$

Squaring (4.35) and equating it with the right hand side of Eq. (4.34) gives after some

4.3. Solitons on the ion time-scale

algebraic manipulations

$$p_\alpha = \frac{V\psi_\alpha \pm R_\alpha}{1-V^2}. \tag{4.38}$$

To decide which sign is correct, we have to take the asymptotic behavior into account. The momentum p_α has to vanish for $\xi \to \infty$ since we are looking for localized solutions. As $R_\alpha|_{\pm\infty} = V$ we identify the correct expression for the momentum as

$$p_\alpha = \frac{1}{\varepsilon_\alpha}\frac{V\psi_\alpha - R_\alpha}{1-V^2}. \tag{4.39}$$

In order to express γ_α in terms of ϕ and $A_\perp$ we start again with the two expressions for the γ_α factor (4.34) and (4.35). First we square (4.35), equate it with the right hand side of (4.34) and then insert (4.39). We get

$$-\frac{R_\alpha^2 + \psi_\alpha^2}{1-V^2} = 1 + \varepsilon_\alpha^2 a^2. \tag{4.40}$$

Inserting this into Eq. (4.35) results in

$$\begin{aligned}\gamma_\alpha^2 &= \frac{\psi_\alpha^2 - R_\alpha^2}{1-V^2} + \frac{(V\psi_\alpha - R_\alpha)^2}{(1-V^2)^2} \\ &= \frac{\psi_\alpha^2 + VR_\alpha^2 - 2VR_\alpha\psi_\alpha}{(1-V^2)^2}.\end{aligned} \tag{4.41}$$

We get an expression for the γ_α factor which is a function of ϕ and $A_\perp$

$$\gamma_\alpha = \frac{\psi_\alpha - VR_\alpha}{1-V^2}. \tag{4.42}$$

To obtain an expression for the densities n_α, start with $v_\alpha = \varepsilon_\alpha \frac{p_\alpha}{\gamma_\alpha}$ and insert (4.39)

4. Stability and dynamics of relativistic 1D solitons

and (4.42), which results in
$$v_\alpha = \frac{V\psi_\alpha/R_\alpha - 1}{\psi_\alpha/R_\alpha - V}. \tag{4.43}$$

We insert this into the expression for n_α which we got by integration of the continuity equation
$$n_\alpha \varepsilon_\alpha \left(\frac{p_\alpha}{\gamma_\alpha} - V\right) = -V, \tag{4.44}$$

and we get
$$n_\alpha = \frac{V^2 - V\psi_\alpha/R_\alpha}{V^2 - 1}. \tag{4.45}$$

We now have expressions for p_α, γ_α and n_α in terms of ϕ and $A_\perp$. Inserting (4.45) into the 1D Laplace equation
$$\frac{\partial^2 \phi}{\partial \xi^2} = n_e - n_i, \tag{4.46}$$

leads to
$$\frac{\partial^2 \phi}{\partial \xi^2} = \frac{V}{1-V^2}\left(\frac{\psi_e}{R_e} - \frac{\psi_i}{R_i}\right). \tag{4.47}$$

The 1D wave equation for the potential $A_\perp$ reads
$$\left(1 - V^2\right)\frac{\partial^2}{\partial \xi^2} A_\perp + 2V \frac{\partial^2}{\partial \xi \partial \tau} A_\perp - \frac{\partial^2}{\partial \tau^2} A_\perp = \left(\varepsilon_i \frac{n_i}{\gamma_i} + \frac{n_e}{\gamma_e}\right) A_\perp$$
$$= A_\perp V \left(\frac{1}{R_e} + \varepsilon_i \frac{1}{R_i}\right). \tag{4.48}$$

We make the ansatz $A_\perp(\xi, \theta) = a(\xi)e^{i\omega\theta}$ for the solitons, where $\theta = t - Vx$. Afterwards we transform the equations into the co-moving frame which is defined by $\tau = t$ and $\xi = x - Vt$. Inserting into Eq. (4.48) gives us the ordinary differential equation for the amplitude of the vector potential
$$\frac{\partial^2 a}{\partial \xi^2} + \omega_0^2 a = \frac{aV}{1-V^2}\left(\frac{1}{R_e} + \varepsilon_i \frac{1}{R_i}\right). \tag{4.49}$$

4.3. Solitons on the ion time-scale

Equations (4.47) and (4.49) describe coupled Langmuir and circular polarized transverse electromagnetic waves. These coupled equations can be solved numerically by standard integration methods like Runge-Kutta-Fehlberg methods.

These equations are reversible under the transformation $\xi \to -\xi$, since $a \to \pm a$ and $\phi \to \phi$. This means that we expect the solutions to be symmetric in ϕ and either symmetric or antisymmetric in a. We restrict ourself to solutions which are single humped in ϕ. To solve the system we choose a fixed value for V and integrate Eq. (4.49) and (4.47) with the initial conditions at $\xi = \xi_1$, $\phi = \phi' = 0$, $a'_1 = \lambda a_1$ for small a_1 and $\lambda = \sqrt{(1+\varepsilon_i)/(1-V^2) - \omega_0^2}$. We will adjust ω_0 until there is a position at which $\phi' = 0$ and $a'_1 = 0$ (for even p) or $\phi' = 0$ and $a_1 = 0$ (for odd p). After having found such a combination of ω_0 and V the calculation will be repeated on a shifted grid in such a way that the maximum of ϕ is located at $\xi = 0$. Once the vector and the scalar potential have been obtained, the other quantities p_α, n_α and γ_α can be calculated from these potentials by Eq. (4.39), (4.45) and (4.42). Figure 4.7 shows three different solitons.

The solitons are categorized by the number $p = 0, 1, 2, \ldots$ of nodes in the vector potential. With mobile ions there is a minimum velocity above which these solitons exist. The $p = 0$ solitons have a continuum in ω_0 for which they exist for a fixed velocity V, this distinguishes them from all other solitons $p = 1, 2, \ldots$ for which there is only a discrete spectrum for every velocity V.

To check the stability of the Bulanov-Farina solitons, we carried out simulations of the system (4.9)-(4.12). For linear stability analysis it is more accurate and convenient to deal with time-independent time-evolution operators (see chapter 3). Hence we transformed out the time dependence of the vector potential due to the phase factor $\exp(i\omega_0(1-V^2)t)$ of the soliton. The associated eigenvalue to the most unstable mode in the linear regime may in general be complex, so this mode might evolve in time as $e^{\Gamma t} \cos(\tilde{\omega} t)$.

4. Stability and dynamics of relativistic 1D solitons

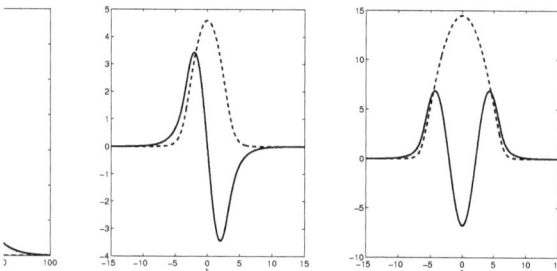

Figure 4.7: Amplitude a of the vector potential (solid line) and electrostatic potential ϕ (dashed line) for solitons with node numbers $p = 0$, $p = 1$ and $p = 2$ (left to right). All solitons have the velocity $V = 0.4$, the frequencies are $\omega_0 = 1.08912$, $\omega_0 = 0.8533$ and $\omega_0 = 0.75766$, respectively.

For solitons with $p = 0$ we did not find any physically relevant growth rates, so we conclude that these solitons are stable. Solitons with node numbers $p = 1, 2, \ldots$ are unstable and show two different types of instabilities. The linear growth rate Γ for these solitons is shown in Fig. 4.8 for different soliton velocities V. In general the growth rate increases as the soliton velocity decreases. For $p = 1$ there is a local minimum of the growth rate at $V \approx 0.7$. This point separates two kinds of instabilities. Perturbations for solitons above this velocity are purely growing by $e^{\Gamma t}$, those below exhibit an additional oscillatory evolution and grow like $e^{\Gamma t} \cos(\tilde{\omega} t)$. This separation between the two types of perturbations can be seen for solitons with higher node numbers p too, but takes place at higher velocities.

Let us discuss the meaning of $\tilde{\omega}$ for the evolution of the perturbation in the laboratory frame. If $\tilde{\omega}$ is zero, the evolution of the perturbation in the laboratory frame at a fixed position is only due to the spatial distribution of the mode in the co-moving frame. For the vector potential the additional phase factor has to be included since to

4.3. Solitons on the ion time-scale

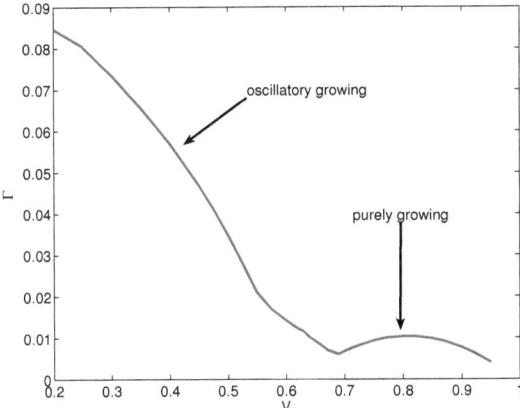

Figure 4.8: Linear growth rate Γ for longitudinal instability of $p = 1$ solitons depending on the soliton velocity V. The minimum at $V \approx 0.7$ separates oscillatory growing perturbations from purely growing ones.

4. Stability and dynamics of relativistic 1D solitons

grow this mode we transformed out the explicit time dependence due to this factor. The oscillations of the vector potential of the perturbation at a fixed position in the laboratory frame are therefore determinded by the soliton frequency ω_0 and the spatial distribution in the co-moving frame.

In the case $\tilde{\omega}$ different from zero the unstable mode oscillates in the co-moving frame and this oscillation introduces additional oscillations in the laboratory frame. This leads to the generation of frequency side bands in the vector potential. Let us again consider the vector potential. First, the spatial distribution in the co-moving frame leads to a temporal evolution in the laboratory frame at a fixed position, these oscillations are then shifted by ω_0 due to the phase factor. Second, the frequency $\tilde{\omega}$ in the co-moving frame leads to side bands $\omega_0 \pm \tilde{\omega}$. These side bands are considered to be an indication of forward Raman scattering by Saxena et al. [76].

In order to verify the linear results and to examine the nonlinear regime of the evolution a nonlinear simulation is done. The initial conditions are the unperturbed soliton plus small amount ϵ of the linear fastest growing perturbation for the soliton. As long as the amount of the perturbation is small, the growth of the perturbation has to be proportional to $e^{\Gamma t}$. The nonlinear terms become important when the perturbation has grown to some reasonable amount or they have been acting for a long time. In the nonlinear regime it might be possible for the perturbation to become saturated.

Figure 4.9 shows the nonlinear evolution of a $p = 1$ soliton with $V = 0.8$ and an added small initial perturbation . Initially the perturbation was 1 % of the soliton. The perturbation excites a growing electrostatic plasma wave behind the soliton. The amplitude of the excited wave is not constant. Figure 4.10 shows the further evolution of the perturbed soliton. At approximately $t = 600$ the excited plasma wave is strong enough to split up the vector potential. A new stable structure is created which is slower than the original soliton. This structure seems to drive the wave-field excitation much more efficiently and very large density gradients are achieved. Between these

4.3. Solitons on the ion time-scale

spikes the electron density becomes more and more depleted and some small part of the vector potential eventually becomes trapped inside these cavities. The steeper the gradients become, the shorter the typical time-scale for the growth of these spikes becomes.

Following this evolution over an even longer time is prohibited due to the fixed finite resolution of the computational grid. Including kinetic effects like a finite temperature might stop the growth of these peaks when the pressure term balances the ponderomotive and the electorstatic force. Since the pressure term is proportional to $T_\alpha \nabla n_\alpha$ (see Eq. (2.28)), we still would have to resolve large gradients in the density for this term to become large enough to balance the radiation pressure, since T_α is for example only 0.1 for a 50keV plasma. Resolving such a gradient would require a vastly finer computational grid, which in return would require too much computational effort to solve the equations on this grid. Methods like adaptive mesh refinement or Godunov methods adapted to Maxwell-fluid models might allow following this process for longer time.

For cold plasma eventually wave-breaking sets in, which results in electric field gradients that become infinite. Figure 4.11 displays the electron density and the electric field for a point in time where wave-breaking is close. The density peaks are already very high and the electric field shows strong gradients.

In the physical picture wave breaking means that particles out of the wave overtake the wave, that is that the peak fluid velocity equals the phase speed of the plasma wave [1, 19]. Wave-breaking is a process where trajectories of adjacent fluid elements would cross; this crossing is associated with a microscopic change of the distribution function. Such effects are not covered by the fluid description, hence this description is no more adequate. Using fluid equations and simulating beyond the point where wave-breaking takes place results in unphysical negative densities. The appearance of these unphysical results from Maxwell-fluid simulations in this situation can be

4. Stability and dynamics of relativistic 1D solitons

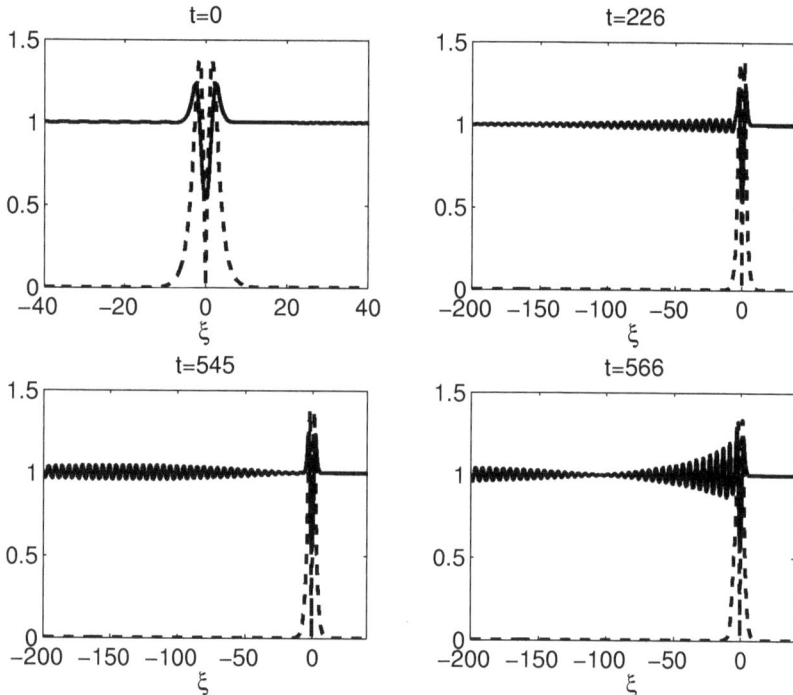

Figure 4.9: Electron density n_e (solid line) and intensity $|a|^2$ (dashed line) of a perturbed $p = 1$ soliton with $V = 0.8$ at times $t = 0$, $t = 226$, $t = 545$ and and $t = 566$ (blue line), respectively. A growing plasma wave is excited behind the soliton, while the vector potential of the soliton does not suffer any major deformation up to $t = 566$. The amount of the initial perturbation is 1% of the unperturbed soliton, the linear growth rate $\Gamma = 0.0104$.

4.3. Solitons on the ion time-scale

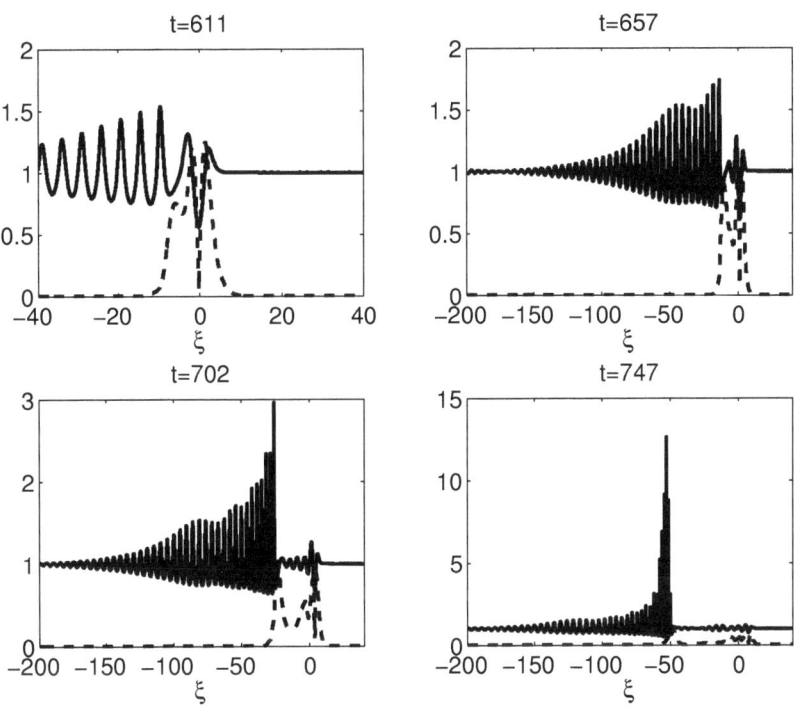

Figure 4.10: Electron density n_e (solid line) and intensity $|a|^2$ (dashed line) of a perturbed $p = 1$ soliton with $V = 0.8$ at times $t = 611$, $t = 657$, $t = 702$ and and $t = 747$ (blue line), respectively. The excited plasma wave becomes strong enough to split up the vector potential. A slower, stable structure is created. Eventually wave-breaking sets in. The amount of the initial perturbation is 1% of the unperturbed soliton, the linear growth rate $\Gamma = 0.0104$.

4. Stability and dynamics of relativistic 1D solitons

understood by an analytical model that is developed in the next chapter.

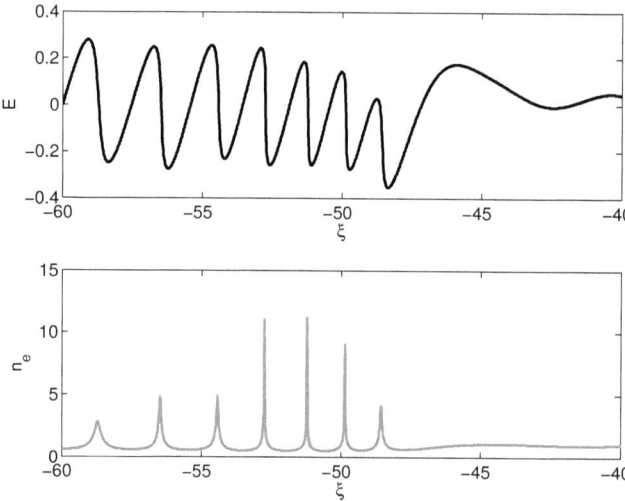

Figure 4.11: Electric field E and electron density n_e at time $t = 747$ of a perturbed $p = 1$ soliton with $V = 0.8$. The amount of the initial perturbation is 1% of the unperturbed soliton. The peaks in the electron density are associated with strong electric field gradients.

4.4 Solitons in warm electron-ion plasma

In some scenarios the effects of a finite temperature can no longer be neglected. The balance between electrostatic and electromagnetic fields inside a soliton can now be altered by the plasma pressure. To study the effects of finite plasma temperature we distinguish two approaches that have been taken so far.

Both approaches are based on a hydrodynamic description for a hot plasma coupled to Maxwell's equations from first principles, namely the conservation laws for particle number and the energy-momentum tensor [57]. Within the first approach the particle distribution function is supposed to be isotropic [57]. The *adiabatic* closure of the fluid equations leads to a entropy conservation for each species. This model has then been employed to study 1D circular polarized non-drifting solitons in overdense electron-positron ($\alpha = e, p$) plasma for arbitrary temperatures $T_{e,p}$. It is found that a finite temperature allows for soliton solutions that would not exist in cold plasma. In the ultrarelativistic regime extremely large concentrations of electromagnetic energy are possible while the temperature profile of the background plasma becomes strongly nonuniform. In the limit of vanishing temperatures the adiabatic model predicts that even a modest field $|A_\perp| \ll 1$ is able to deplete the plasma density inside the center of the soliton up to full expulsion. It is due to the same mass of electrons and positrons that no charge separation is expected, hence a stationary solution has no electrostatic potential. Since there is no charge separation, even solitons with very low amplitudes are able to expel all density from their inside. A similar analysis has been done in Ref. [6] using the slowly varying envelope approximation in an underdense electron-positron plasma.

Treatment of an electron-ion plasma is more complicated than for an electron-positron plasma, since in general electrostatic potentials are present. In situations where the particle thermal spread in transverse direction(to the radiation direction)

4. Stability and dynamics of relativistic 1D solitons

is small compared to the spread in longitudinal direction the particle distribution function is highly anisotropic. The anisotropic case is the more physically relevant one in terms of laser-plasma interaction. In Ref. [58] a two-fluid electron-ion plasma model is derived under the assumption of such an anisotropic distribution function for both species. The plasma is supposed to have a constant parallel temperature due to fast quasithermalization in longitudinal direction. In transversal direction the distribution function is highly anisotropic due to the motion of the particles in the laser fields. The transversal thermal spread will be negligible compared to the motion under the influence of the electromagnetic fields. Thus, the particle distribution function f_α is assumed to be beam-like

$$f_\alpha(W_\alpha, \mathbf{P}_{\alpha\perp}) = \frac{n_{0\alpha}}{2K_1(T_\alpha^{-1})} \delta(\mathbf{P}_{\alpha\perp}) \exp\left(-\frac{W_\alpha}{\varepsilon_\alpha T_\alpha}\right), \tag{4.50}$$

where $n_{\alpha 0}$ is the background density of species α, W_α is the total energy of a particle, T_α the ratio of thermal energy to the rest energy and K_1 the modified Bessel function of first order.

The δ-function assures the conservation of the transversal component of the generalized momentum. The total particle energy at position $\mathbf{r}$ is given by

$$W_\alpha(\mathbf{r}, t, p_\alpha) = \gamma_s + q_s \phi(\mathbf{r}, t), \tag{4.51}$$

where the dependence of the parallel momentum p_α is introduced by the relativistic factor

$$\gamma_\alpha = \left(1 + \varepsilon_\alpha^2 p_\alpha^2\right)^{1/2}. \tag{4.52}$$

The generalized momentum is determined by

$$\mathbf{P}_\alpha(\mathbf{r}, t) = \mathbf{p}_\alpha + q_\alpha \mathbf{A}(\mathbf{r}, t). \tag{4.53}$$

4.4. Solitons in warm electron-ion plasma

The distribution function (4.50) is an exact solution of the relativistic Vlasov equation

$$\frac{\partial}{\partial t}f_\alpha + \frac{\mathbf{p}_\alpha}{\gamma_\alpha}\cdot\nabla f_\alpha + q_\alpha\left[\mathbf{E}(\mathbf{r},t) + \frac{\mathbf{p}_\alpha}{\gamma_\alpha}\times\mathbf{B}(\mathbf{r},t)\right]\cdot\frac{\partial f_\alpha}{\partial \mathbf{p}_\alpha} = 0. \tag{4.54}$$

Inserting (4.50) into (4.54) results in

$$\frac{1}{T_\alpha}\delta\left(\mathbf{P}_{\alpha\perp}\right)\left(\frac{\partial}{\partial t}\phi - \frac{\partial}{\partial t}\mathbf{A}\cdot\frac{\mathbf{p}_\alpha}{\gamma_\alpha}\right) + \left[\nabla\phi - (\nabla\mathbf{A})\cdot\frac{\mathbf{p}_\alpha}{\gamma_\alpha}\right]\cdot\frac{\partial}{\partial \mathbf{P}_{\alpha\perp}}[\delta(\mathbf{P}_{\alpha\perp})] = 0. \tag{4.55}$$

If we suppose one-dimensional geometry, where all quantities only depend on one spatial coordinate (assume x), assume circular polarization and use the the conservation of $\mathbf{P}_{\alpha\perp}$, Eq. (4.55) and therefore (4.54) are exactly satisfied by an electromagnetic distribution with stationary energy. Therefore (4.50) is an exact solution of the 1D kinetic equation and can be used to calculate the consistent charge and current density distributions. These distributions are the field sources in Maxwell's equations and are given by

$$n_\alpha(\mathbf{r},t) = n_{\alpha 0}\frac{K_1[\gamma_{\alpha\perp}(\varepsilon_\alpha T_\alpha)^{-1}]}{K_1[(\varepsilon_\alpha T_\alpha)^{-1}]}\gamma_{\alpha\perp}\exp\left(-\frac{q_\alpha\phi}{\varepsilon_\alpha T_\alpha}\right), \tag{4.56}$$

$$\mathbf{j}_{\alpha\perp}(\mathbf{r},t) = -q_\alpha n_{\alpha 0}\frac{K_0[\gamma_{\alpha\perp}(\varepsilon_\alpha T_\alpha)^{-1}]}{K_1[(\varepsilon_\alpha T_\alpha)^{-1}]}\mathbf{A}_\perp\exp\left(-\frac{q_\alpha\phi}{\varepsilon_\alpha T_\alpha}\right), \tag{4.57}$$

where we introduced $\gamma_{\alpha\perp} = \sqrt{1+\varepsilon_\alpha^2\mathbf{A}_\perp^2}$ and $K_n(x)$ is the modified Bessel function of order n. As in previous chapters we combine the two transversal components of the vector potential into one complex quantity $A_\perp = A_y + iA_x$. Due to 1D geometry and Coulomb-gauge we have $A_x = 0$. By inserting the current densities into the 1D wave

4. Stability and dynamics of relativistic 1D solitons

equation for vector potential we get

$$\frac{\partial^2}{\partial x^2}A_\perp - \frac{\partial^2}{\partial t^2}A_\perp = A_\perp \left\{ \frac{K_0(\sqrt{1+A_\perp^2}T_e^{-1})}{K_1(T_e^{-1})} \exp\left(\frac{\phi}{T_e}\right) \right.$$
$$\left. +\varepsilon_i Z \frac{K_0\left[\sqrt{1+\varepsilon_i^2 Z^2 A_\perp^2}(\varepsilon_i T_i)^{-1}\right]}{K_1[(\varepsilon_i T_i)^{-1}]} \exp\left(-\frac{Z\phi}{T_i}\right) \right\} \quad (4.58)$$

with ion charge state Z.

We are looking for nondrifting localized solutions of the form $A_\perp(x,t) = a(x)e^{i\omega_0 t}$, with frequency ω_0. By inserting this ansatz into Eq. (4.58) we get

$$\frac{\partial^2}{\partial x^2}a + \omega_0^2 a = a \left\{ \frac{K_0(\sqrt{1+A_\perp^2}T_e^{-1})}{K_1(T_e^{-1})} \exp\left(\frac{\phi}{T_e}\right) \right.$$
$$\left. +\varepsilon_i Z \frac{K_0\left[\sqrt{1+\varepsilon_i^2 Z^2 A_\perp^2}(\varepsilon_i T_i)^{-1}\right]}{K_1[(\varepsilon_i T_i)^{-1}]} \exp\left(-\frac{Z\phi}{T_i}\right) \right\}. \quad (4.59)$$

The equation for the electrostatic potential reads

$$\frac{\partial^2}{\partial x^2}\phi = \sqrt{1+a^2}\frac{K_1(\sqrt{1+a^2}T_e^{-1})}{K_1(T_e^{-1})} \exp\left(\frac{\phi}{T_e}\right)$$
$$-\sqrt{1+\varepsilon_i^2 Z^2 a^2}\frac{K_1\left[\sqrt{1+\varepsilon_i^2 Z^2 a^2}(\varepsilon_i T_i)^{-1}\right]}{K_1[(\varepsilon_i T_i)^{-1}]} \exp\left(-\frac{Z\phi}{T_i}\right). \quad (4.60)$$

Equations (4.59) and (4.60) are a closed set of 1D equations for relativistic electromagnetic fields interacting in a warm, two-component plasma. The macroscopic state has been derived from a solution of the kinetic Vlasov equation.

Let us study solutions to Eq. (4.59) and (4.59) by assuming quasi-neutrality, i.e. $(n_e - n_i)/n_0 \ll 1$, where n_0 is the background density (assuming $Z = 1$). In this case Eq. (4.60) can be put to zero and the right hand side gives us an expression for the

4.4. Solitons in warm electron-ion plasma

electrostatic potential in terms of the vector potential amplitude, i.e.

$$\phi(a^2;\varepsilon_i,T_e,T_i) = \left(\frac{1}{T_e}+\frac{1}{T_i}\right)^{-1}\left\{\frac{1}{2}\ln\frac{1+\varepsilon_i^2 a^2}{1+a^2}+\ln\frac{K_1(T_e^{-1})K_1[\sqrt{1+\varepsilon_i^2 a^2}(\varepsilon_i T_i)^{-1}]}{K_1[(\varepsilon_i T_i)^{-1}]K_1[\sqrt{1+a^2}(T_e)^{-1}]}\right\}.$$
(4.61)

Expression (4.61) can be inserted into the right hand side of (4.59) to eliminate ϕ. The result is a second-order ordinary differential equation for $a(x)$,

$$\frac{\partial^2}{\partial x^2}a + \omega_0^2 a = G(a) \tag{4.62}$$

where

$$G(a) = a\left\{\frac{K_0(\gamma_{e\perp}/T_e)}{K_1(1/T_e)}\left[\frac{\gamma_{i\perp}K_1(1/T_e)K_1(\gamma_{i\perp}/\varepsilon_i T_i)}{\gamma_{e\perp}K_1(1/\varepsilon_i T_i)K_1(\gamma_{e\perp}/T_e)}\right]^{\lambda}\right.$$
$$\left.+\varepsilon_i\frac{K_0(\gamma_{i\perp}/\varepsilon_i T_i)}{K_1(1/\varepsilon_i T_i)}\left[\frac{\gamma_{i\perp}K_1(1/T_e)K_1(\gamma_{i\perp}/\varepsilon_i T_i)}{\gamma_{e\perp}K_1(1/\varepsilon_i T_i)K_1(\gamma_{e\perp}/T_e)}\right]^{\lambda-1}\right\}, \quad (4.63)$$

with $\lambda = T_i/(T_i+T_e)$.

For various values of the parameters ω, T_e, and T_i we have determined the soliton solutions, in agreement with Ref. [58]. In Figure 4.12 a soliton is depicted for the parameters $T_e = T_i = 30$ and $\omega_0 = 0.1$. Eventhough the electrostatic potential has a large amplitude, the second derivative of ϕ is still small, hence the assumption of quasi-neutrality is still met to a good approximation.

The stability of solitons in warm plasma is of importance since kinetic effects may be of importance for the balance between ponderomotive and electrostatic force. The evolution of possible perturbations is affected by additional pressure terms, which could dampen oscillations in the electrostatic potential.

In order to perform a stability analysis in terms of small perturbations in the am-

4. Stability and dynamics of relativistic 1D solitons

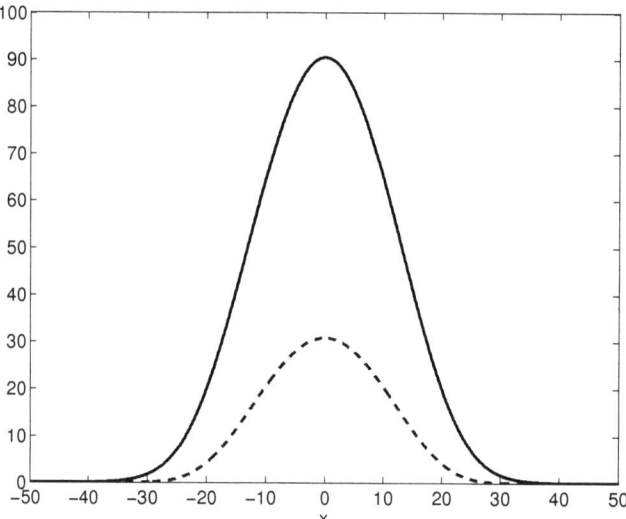

Figure 4.12: Amplitude of vector potential a (solid line) and electrostatic potential ϕ (dashed line) for $T_e = T_i = 30$ and $\omega_0 = 0.1$. The solutions where obtained by numerical solution of Eq. (4.62) and (4.61)

4.4. Solitons in warm electron-ion plasma

plitude of the vector potential we have to linearize Eq. (4.62) about an unperturbed solution. The right hand side of the linearized equation is given by

$$G' = \frac{\lambda B^{\lambda-1} K_0(\gamma_{e\perp}/T_e) B'}{K_1(1/T_e)} + \frac{(\lambda-1)\varepsilon_i B^{\lambda-2} K_0(\gamma_{i\perp}/\varepsilon_i T_i) B'}{K_1(1/\varepsilon_i T_i)} - \frac{B^\lambda K_1(\gamma_{e\perp}/T_e)}{K_1(1/T_e) T_e \gamma_{e\perp}} - \frac{B^{\lambda-1} K_1(\gamma_{i\perp}\varepsilon_i T_i)\varepsilon^2}{K_1(1/\varepsilon_i T_i)\gamma_{i\perp} T_i}, \quad (4.64)$$

where

$$B' = \left\{ K_1(1/T_e) \left[-\frac{\varepsilon_i}{\gamma_{e\perp}} K_1(\gamma_{i\perp}/\varepsilon_i T_i) T_i \gamma_{i\perp} \left(K_1(\gamma_{e\perp}/T_e) T_e - \frac{\gamma_{e\perp}}{2} \left(K_0(\gamma_{e\perp}/T_e) + K_2(\gamma_{e\perp}/T_e) \right) \right) \right] \right.$$
$$\left. + \frac{K_1(\gamma_{e\perp}/T_e)\varepsilon_i^2}{\gamma_{i\perp}} \gamma_{e\perp} T_e \left[\varepsilon_i K_1(\gamma_{i\perp}/\varepsilon_i T_i) T_i - \frac{\gamma_{i\perp}}{2} \left(K_0(\gamma_{i\perp}/\varepsilon_i T_i) + K_2(\gamma_{i\perp}/\varepsilon_i T_i) \right) \right] \right\} /$$
$$\left(\varepsilon_i K_1(1/\varepsilon_i T_i) K_1(\gamma_{e\perp}/T_e)^2 T_e T_i \gamma_{e\perp}^2 \right). \quad (4.65)$$

The linear evolution of a perturbation a_1 in the amplitude of the vector potential $a = a_0 + \epsilon a_1$ (ϵ is a smallness parameter, a_0 is the unperturbed solution) is therefore determined by the equation

$$\frac{\partial^2}{\partial t^2} a_1 + \omega_0^2 a_1 = G'(a_0). \quad (4.66)$$

Figure 4.13 shows the time-evolution of the maximum of the perturbation amplitude. We recognize a *linear* growth of the perturbation. It corresponds to a solution $i\,\delta\omega\,t\,a_s(x)\,e^{i\omega t}$ (the so called phasor-mode, see Sec. 3.2) of Eq. (4.66). The interpretation as a phasor mode, and not as an exponential instability, is supported by the following comparison. The structure of the linearly growing mode follows from numerics. When compared to $|i\,a_s(x)\,e^{i\omega t}|$, see Fig. 4.14, we recognize perfect agreement. The stability of the solitons in warm plasmas also follows from the Q-theorem [60, 49],

4. Stability and dynamics of relativistic 1D solitons

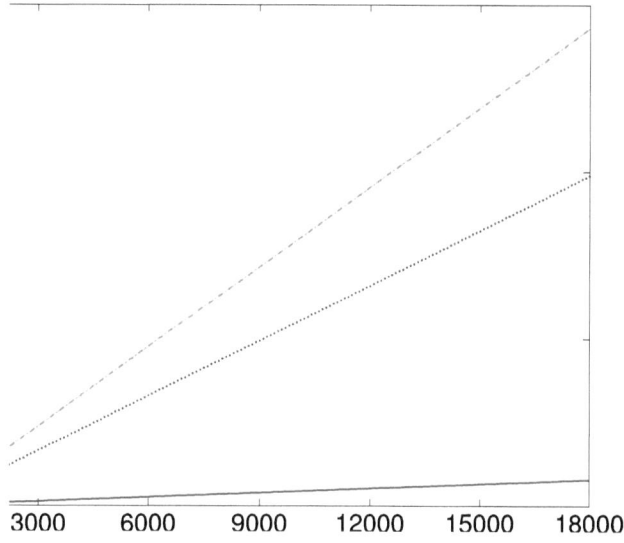

Figure 4.13: Time-evolution of the maximum $|a|_{max}$ of the perturbation amplitude from numerical integration of the linearized version of equation (4.62) with $\omega = 0.1$. The lines corresponds to $T_e = T_i = 0.002$ (solid), $T_e = T_i = 0.1$ (dotted), and $T_e = T_i = 10$ (dash–dotted), respectively.

4.4. Solitons in warm electron-ion plasma

see Appendix B.1. The solution $a(x,t) = a_s(x)e^{i\omega t}$ is stable provided

$$Q = i \int dx \left(a \frac{\partial a^*}{\partial t} - a^* \frac{\partial a}{\partial t} \right) \tag{4.67}$$

is a monotonically decreasing function of ω. Figure 4.15 demonstrates that behavior.

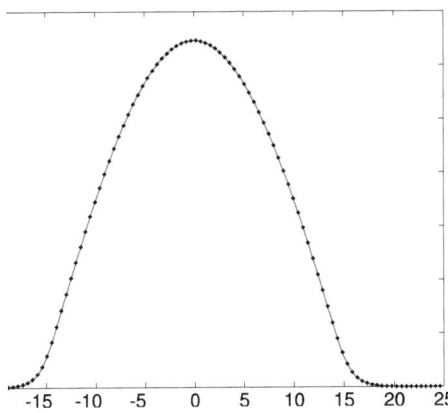

Figure 4.14: Form of the linearly growing perturbation (solid line) for a soliton with $T_e = T_i = 0.1$ and $\omega = 0.1$ compared to $|i\, a_s(x)\, e^{i\omega t}|$ (crosses). Note that the amplitudes have been adjusted to demonstrate the complete agreement.

4. Stability and dynamics of relativistic 1D solitons

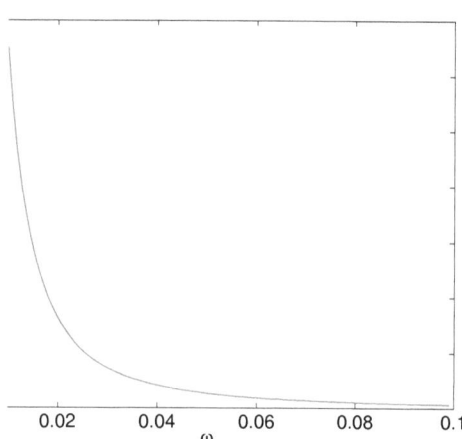

Figure 4.15: Evaluation of the Q-theorem according to Refs. [49]. Shown is Q versus ω, demonstrating stability of the solitons in warm plasmas for $T_e = T_i = 10$.

5 Relativistic wave-breaking in cold plasma

In the preceding chapter we observed wave-breaking processes as part of the unstable dynamics of 1D solitons. The nonlinear evolution of the perturbed solitons showed density modulations that are left behind the soliton and begin to form very sharp density peaks. This leads to strong gradients in the electrostatic field. To gain further understanding of this process and allow to distinguish physical from numerical effects, an appropriate model for this scenario has to be formulated.

Wake-fields are electrostatic waves that may be created by the vector potential of a laser pulse. In cold plasma the wake-field stays at a fixed position and has only a phase velocity (which is determined by the group velocity of the laser). In this chapter we will discuss the creation and the stability of wake-fields in cold plasma. We will derive a condition that describes the wake-field generation and afterwards apply a Lagrangian coordinate description of the plasma to generalize known wave-breaking criteria to the relativistic regime.

5.1 Laser wakefields for particle acceleration

Before we start to study the physics of the wake-excitation process and the dynamics of the wake-field, let us discuss the importance of wake-fields for particle accelerators

5. Relativistic wave-breaking in cold plasma

and some results form literature.

The stability of wake-fields is of very general interest, since they are a basic constituent of laser-plasma accelerators for charged particles. Conventional particle accelerators are limited to accelerating fields of the order of roughly 100 MV/m , partly because of the breakdown occurring on the walls of the structure. Ionized plasmas on the other hand can sustain electric fields up to several 100 GV/m. If these fields could be used efficiently it would give access to small-scale high-energy accelerators.

A laser pulse propagating through an underdense plasma excites a plasma wave oscillating at the plasma frequency ω_{pe}. The phase velocity of this oscillation is set by the group velocity of the laser $v_{ph}^{wake} = v_g^{laser} = c\sqrt{1 - \omega_{pe}^2/\omega_0^2}$, where ω_0 is the laser frequency. This longitudinal field may exceed the cold nonrelativistic wave breaking field [19] $E_0 = cm_e\omega_p/e$ or

$$E_0(\text{V/m}) \simeq 96\sqrt{n_0(\text{cm}^{-3})}. \tag{5.1}$$

For example, a plasma density of $n_0 = 10^{18}\text{cm}^{-3}$ yields $E_0 \simeq 100$ GV/m. This is a factor of three orders of magnitude in comparison to conventional accelerators, and so should allow for much shorter acceleration distances.

Besides very large acceleration gradients, wake-field-based accelerators might produce very short electron bunches. The acceleration length is approximately the plasma wavelength $\lambda_{pe} = 2\pi c/\omega_{pe} = 2\pi/k_p$ or

$$\lambda_p(\mu\text{m}) \simeq 3.3 \times 10^{10}/\sqrt{n_0(\text{cm}^{-3})}. \tag{5.2}$$

A high-quality electron bunch would have a bunch duration of $\tau_b < \lambda_p/c$, i.e. a duration $\tau_b < 100$ fs for $n_0 = 10^{18}\text{cm}^{-3}$ [4, 52].

The laser wake-field acceleration (LWFA) scheme was first proposed by Tajima and

5.1. Laser wakefields for particle acceleration

Dawson [84]. Ten years later, when the laser power made huge step towards higher intensities by the invention of the chirped pulse amplification method [83], it was refined and extended [31, 82].

As an intense laser pulse propagates through an underdense plasma, $\omega_0/\omega_{pe} \gg 1$, the ponderomotive force associated with the laser pulse envelope, $F_p \sim \nabla A_\perp^2$, expels electrons from the region of the laser pulse. Let the direction of propagation be x. If the length scale L_x of the axial gradient of the the laser potential is roughly equal to the plasma wave length, $L_x \sim \lambda_p$, the ponderomotive force excites large amplitude plasma waves which are called wake-field. So the wake-field is driven most efficiently when $L \approx \lambda_p$. The precise amplitude of the wake depends on the axial profile of the pulse, analytical models are available for sine and Gaussian pulses only.

In the linear regime the electrical field is of the form $E_x = E_{\max} \sin[\omega_{pe}(x/v_p - t)]$ with phase velocity $v_p \simeq c$. The maximum wake amplitude can be estimated from the Poisson equation $\nabla \cdot \mathbf{E} = 4\pi e(n_0 - n_e)$. The most simple estimate is that all electrons oscillate with a wave number $k_p = \omega_{pe}/c$. This gives $(\omega_{pe}/c)E_{\max} = 4\pi e n_0$, or $E_{\max} = E_0$, where $E_0 = c m_e \omega_{pe}/e$ is the cold nonrelativistic wave-breaking field [19].

For nonlinear plasma waves it is possible to exceed the value E_0. Using a nonlinear, relativistic, cold fluid approximation in 1D, it was shown [1] that the maximum electrical field is given by

$$E_{\text{WB}} = \sqrt{2}(\gamma_p - 1)^{1/2} E_0 \,, \tag{5.3}$$

which is referred to as the cold relativistic wave-breaking field. The relativistic Lorentz factor is given by $\gamma_p = \sqrt{1 - \beta_p^2}$ with $\beta_p = v_p/c$. Thus it is possible to exceed the nonrelativistic maximum value by orders of magnitude. A typical value for γ_p would be 100 for a 1 μm laser, if we consider a plasma with $n_0 = 10^{17}$ this leads to $E_{WB} = 24 E_0$.

Using a 1D cold relativistic fluid model [5] (which applies for broad laser pulses

5. Relativistic wave-breaking in cold plasma

where $k_p r_\perp \gg 1$, with $r_\perp$ as characteristic radial dimension), the Poisson equation can be rewritten as

$$\frac{d^2\phi}{dx^2} = \gamma_p^2 \left(\beta_p \left[1 - \frac{1+A_\perp^2}{\gamma_p^2(1+\phi)^2} \right]^{-1/2} - 1 \right). \tag{5.4}$$

The maximum and minimum values of the scalar potential, denoted by ϕ_m, are determined by [22]

$$\phi_m = \left(\frac{1}{2}\frac{E_{max}}{E_0}\right)^2 \pm \beta_p \left[(1+\left(\frac{1}{2}\frac{E_{max}}{E_0}\right)^2 - 1 \right]^{-1/2}. \tag{5.5}$$

When the electric field approaches the breaking field E_{WB} Eq. (5.5) implies $1+\phi \to 1/\gamma_p$, which in combination with the Poisson equation (5.4) leads to $n \to \infty$. This is the point where the cold fluid description breaks down. It is only valid for wave-breaking in the limit $\gamma_p \beta_p \ll 1$, with $c\beta_p$ as the thermal velocity spread of the electrons. Thermal effects will broaden the velocity distribution in a warm plasma and reduce the maximum plasma wave amplitude.

Several works dealt with the problem of correcting the cold nonrelativistic field E_0 by using warm fluid models in different limits [18, 74, 38]. The extension to warm relativistic plasmas is discussed in Refs. [78, 80, 86, 79].

All the above mentioned results are derived from 1D models. In 3D only results from numerical simulations are available and analytical expressions for maximum field amplitudes are still missing. Some 2D Particle-In-Cell (PIC) simulations have shown fields of the order of E_0 [15, 20], and 2D axisymmetric nonlinear fluid simulations [43] showed wave amplitudes larger that E_0.

For the wake-field, it is not only possible to break in longitudinal direction, but in transversal direction, too. This 2D wave-breaking is due to the curvature of the phase fronts of the plasma wave [16].

Trapping of particles in plasma waves, which are then accelerated by the wave is

5.1. Laser wakefields for particle acceleration

a very important question for LWFA. Wave-breaking is of central interest, since it provides a mechanism for particle injection into the wake. Particles from the background will become trapped in the wave and accelerated. The predicted conversion efficiency from laser energy into electron beam energy is quite high in this scenario. In experiments where the wave was driven beyond the breaking threshold, large numbers of accelerated electrons have been observed [67, 17, 85]. Particle injection done with external beams and subsequent acceleration has been demonstrated experimentally [3].

To examine the dynamics of wave-breaking and not only give a maximum sustainable electric field amplitude a one-dimensional nonlinear relativistic second-order differential equation for the electron fluid velocity was derived in Ref. [21] in Lagrangian coordinates. This formulation allows to analyze the dynamics until wave-breaking. The Lagrangian analysis, combined with an appropriate numerical integration of the second-order differential equation, gave further insight into the wave-breaking process. We will take up the Lagrangian description to derive analytical criteria for wave-breaking. The analysis is based on Dawson's picture [19] that longitudinal wave-breaking in a cold plasma occurs when elements of the plasma electron fluid that started out in different positions overtake each other. From literature [86] we cite that for both, non-relativistic and relativistic plasmas, this overtaking happens when the peak fluid velocity equals the phase speed of the plasma wave [1, 19]. The existing wave-breaking criteria and thresholds have to be generalized with special emphasis on relativistic and inhomogeneous effects. In addition we shall not restrict ourselves to wave-breaking on the first oscillation. Everything in this chapter will be treated in 1D geometry, that is all quantities only depend on one spatial coordinate x.

5. Relativistic wave-breaking in cold plasma

5.2 Wake-field excitation

The electrostatic wake-field is driven by the vector potential of the laser pulse [31]. For relatively small amplitudes we may use the Poisson equation for the scalar potential ϕ,

$$\frac{\partial^2 \phi}{\partial x^2} = \delta n_e \,, \tag{5.6}$$

together with the linearized particle density conservation law

$$\frac{\partial \delta n_e}{\partial t} + \frac{\partial}{\partial x}\left(n_0 v_\parallel\right) \approx 0 \,. \tag{5.7}$$

Here, n_0 is the zeroth order background particle density. We have assumed a 1D model with fixed ions. The electron density deviation is $\delta n_e = n_e - n_0$. In the non-relativistic limit, the parallel (to the laser propagation direction) electron velocity component $v_\parallel$ is, to lowest order, determined by

$$\frac{\partial v_\parallel}{\partial t} \approx \frac{\partial \phi}{\partial x} - v_\perp B \,. \tag{5.8}$$

The perpendicular component of the electron momentum leads to zeroth order (i.e. in the non-relativistic limit) to the quiver momentum $p_\perp \approx A_\perp$, where $A_\perp$ is the transversal laser vector potential. The space dependence of $p_\perp$ follows (in the perpendicular momentum equation) from the balance of the inertia term with the Lorentz force, i.e.

$$\frac{\partial p_\perp}{\partial x} \approx B \,, \tag{5.9}$$

where $p_\perp = \gamma v_\perp \approx v_\perp$ in the weakly relativistic case for $\gamma \equiv \gamma_e = \sqrt{1 + p_\perp^2} \approx 1$.

Combining these equations and neglecting higher nonlinearities leads to

$$\frac{\partial^2 \delta n_e}{\partial t^2} + \delta n \approx \frac{n_0}{2} \frac{\partial^2 v_\perp^2}{\partial x^2} \,. \tag{5.10}$$

5.2. Wake-field excitation

Equation (5.10) is the equation for the driven density response.

Assuming

$$v_\perp = \frac{1}{2}\left[A_\perp(\xi)e^{-i\omega t + ikx} + A_\perp^*(\xi)e^{i\omega t - ikx}\right], \tag{5.11}$$

we recognize that different harmonics appear on the right-hand-side. For the lowest harmonic density reaction, the solution is

$$\frac{\delta n_e}{n_0} = \frac{1}{4v_g^2}\left\{|A_\perp|^2 - k_p \int_\infty^\xi d\xi'\, |A_\perp(\xi')|^2 \sin[k_p(\xi - \xi')]\right\}, \tag{5.12}$$

where $\xi = x - v_g t$ and $k_p = \omega_{pe}/v_g$. The group velocity v_g is that of the laser pulse.

Different limits can be derived from this solution. First, for narrow laser pulses, e.g. in the simple approximation (pulse width $l \to 0$)

$$|A_\perp(\xi')|^2 = A^2 \delta(\xi'), \tag{5.13}$$

we obtain for finite wave-numbers k_p the constant envelope wake-field

$$\frac{\delta n_e}{n_0} = -\frac{k_p}{4v_g^2} A^2 \sin(k_p x - \omega_{pe} t). \tag{5.14}$$

In general, the wake-field generation vanishes when $k_p l \gg 1$, where l is the width of the driver. Significant contributions occur for

$$k_p l \sim \mathcal{O}(1). \tag{5.15}$$

This is the case we will consider in more detail in the following Section.

On the other hand, in the limit $k_p \to 0$, v_g finite, we obtain

$$\frac{\delta n_e}{n_0} = \frac{1}{4v_g^2}|A_\perp|^2, \tag{5.16}$$

5. Relativistic wave-breaking in cold plasma

i.e. a density hump instead of the ponderomotive driven density depression.

Finally, for $v_g \approx \frac{k}{\omega} \to 0$, ω_{pe} finite, $k_p \to \infty$ we can recover from (5.12) the ponderomotive density depression of a moving laser pulse [63, 32]. In this limit, a first integration by parts in (5.12) leads to

$$\frac{\delta n_e}{n_0} = \frac{1}{4v_g^2} \int_\infty^\xi d\xi' \frac{d|A_\perp(\xi')|^2}{d\xi'} \cos[k_p(\xi - \xi')] . \tag{5.17}$$

Continuing with an additional integration by parts, we obtain from here

$$\frac{\delta n_e}{n_0} = \frac{1}{4v_g^2} \frac{1}{k_p} \int_\infty^\xi d\xi' \frac{d^2|A_\perp(\xi')|^2}{d\xi'^2} \sin[k_p(\xi - \xi')] . \tag{5.18}$$

The next integration by parts gives

$$\begin{aligned}\frac{\delta n_e}{n_0} &= \frac{1}{4v_g^2 k_p^2} \left\{ \frac{d^2|A_\perp|^2}{d\xi^2} - \int_\infty^\xi d\xi' \frac{d^3|A_\perp(\xi')|^2}{d\xi'^3} \cos[k_p(\xi - \xi')] \right\} \\ &\approx \frac{1}{4v_g^2 k_p^2} \frac{d^2|A_\perp|^2}{d\xi^2} + \mathcal{O}\left(\frac{1}{k_p}\right) .\end{aligned} \tag{5.19}$$

This is the result [32, 53] which has been used for soliton investigations.

5.2.1 Numerical simulations and breaking of wake-fields

We first simulated laser pulse propagation in an underdense plasma by solving the set of equations for a plasma with fixed ion background density. The equations result from Eq. (4.3)-(4.6), for a frame of reference with $V = 0$ and constant ion density. We will again combine the transversal components of the vector potential into one

5.2. Wake-field excitation

complex quantity $A_\perp = A_y + \mathrm{i} A_z$. The equations read

$$\frac{\partial^2}{\partial x^2} A_\perp - 1\frac{\partial^2}{\partial t^2} A_\perp = \frac{n_e}{\gamma_e} A_\perp , \qquad (5.20)$$

$$\frac{\partial^2}{\partial x^2} \phi = n_e - 1 , \qquad (5.21)$$

$$\frac{\partial}{\partial t} n_e + \frac{\partial}{\partial x} n_e \frac{p_e}{\gamma_e} = 0 , \qquad (5.22)$$

$$\frac{\partial}{\partial t} p_e = \frac{\partial}{\partial x} (\phi - \gamma_e) \qquad (5.23)$$

where

$$\gamma_e = \sqrt{1 + |A_\perp|^2 + p_e^2} . \qquad (5.24)$$

So far, ion dynamics is neglected here, but will be included later.

Fig. 5.1 shows a snapshot of a simulation of Eq. (5.20)-(5.23). The initial pulse shape of the circular polarized laser pulse is Gaussian with an amplitude of $A_{\perp 0} = 0.15$, the laser frequency is $\omega_0 = 1.4$ and FWHM=40.

The numerical results show the following behavior. The electromagnetic pulse changes periodically its shape. When its width decreases, a significant wake-field can be generated. On the other hand, when the pulse is broad, no wake-field appears. The wake-field has a finite wave-number $k = k_p$, a frequency $\omega = \omega_{pe}$, and a phase velocity

$$\frac{\omega}{k} \approx \frac{\omega_{pe}}{k_p} \approx v_g . \qquad (5.25)$$

The group velocity v_g of the laser pulse is close to the speed of light. The numerical results are in accordance with the analytical predictions. Especially, the estimate of the wake-field localization via the wake-field excitation threshold (5.15), determining the width of the wake-field envelope in terms of the time-dependent pulse width, is quite good. We would like to emphasize that, due to the cold plasma approximation,

5. Relativistic wave-breaking in cold plasma

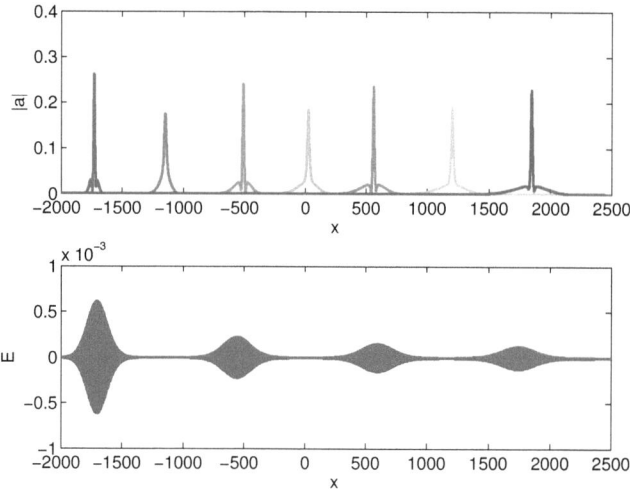

Figure 5.1: The magnitude of the normalized vector potential of an initially Gaussian shaped laser pulse with circular polarization is shown in the top graph at different times when propagating in an under-dense plasma. Equations (5.20) - (5.21) have been solved numerically. The maximum initial amplitude is $A_{\perp 0} = 0.15$, the laser frequency is $\omega_0 = 1.4$, and $FWHM = 40$ for the initial Gaussian. From left to right profiles are plotted at times $t = 700$, $t = 1525$, $t = 2440$, $t = 3200$, $t = 4000$, $t = 4880$ and $t = 5800$. Bottom: Wake-field for $t = 6400$ created by the pulse, corresponding to the different shapes of the laser pulse shown at the top. The spots are created at times where the compression of the laser pulse is at maximum.

the decoupled, localized wake-field is not propagating in space.

We now analyze in more detail the dynamics of the wake-field being left behind the laser pulse. Figure 5.2 presents a typical example. The numerical simulations show the breaking of the wake-field.

A small amplitude localized wake-field shows breaking after many plasma periods. The breaking of the wave is easily diagnosed by the rapid growth of the local electron

5.3. Wave-breaking calculations in Lagrangian coordinates

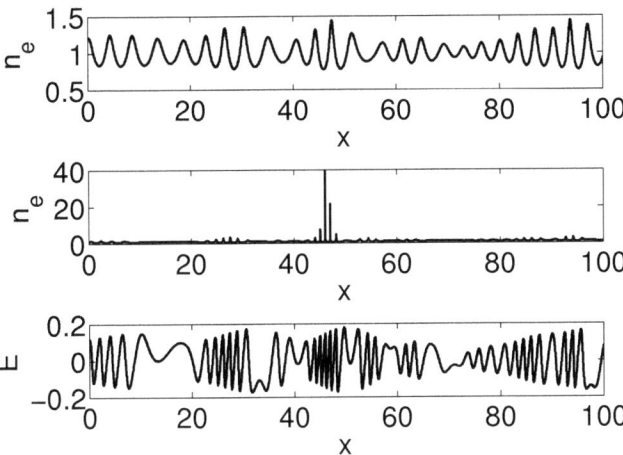

Figure 5.2: Electron density n_e [at $t = 1050$ (upper graph) and $t = 6300$ (middle)] and electric field E at $t = 6300$ (bottom), respectively, from numerical simulation of (5.20) - (5.23). The initial values correspond to an initially Gaussian shaped laser pulse with circular polarization. The maximum initial amplitude is $A_{\perp 0} = 0.32$, the laser frequency is $\omega_0 = 1.9$, the wave-number is $k=1.62$, and $FWHM = 20$ for the initial Gaussian.

density. In the next section we shall analyze the breaking phenomenon analytically.

5.3 Wave-breaking calculations in Lagrangian coordinates

In this section we analyze the observed wave-breaking and derive analytical criteria using a Lagrangian formulation.

We start with a cold fluid description and Lagrange coordinates in one space dimension. The fluid element was at x_0 at $t = 0$ and is at x_L at t, i.e. the La-

5. Relativistic wave-breaking in cold plasma

grange coordinate is $x_L = x_L(x_0, t)$ with $x_L(x_0, t = 0) = x_0$. Defining also Lagrangian electron momentum $p_L = p_L(x_0, t) = p(x_L, t)$, the Lagrange electrostatic field $E_L = E_L(x_0, t) = E(x_L, t)$, and electron density $n_L = n_L(x_0, t) = n(x_L, t)$, the ion density N is assumed to be fixed.

Maxwell's equations (we consider electrostatic wake-fields and neglect the magnetic fields) lead to

$$\frac{\partial E_L}{\partial t} = n_L v_L . \tag{5.26}$$

The total time-derivative of E_L is

$$\frac{dE_L}{dt} = \frac{\partial E}{\partial x_L}\frac{dx_L}{dt} + \frac{\partial E}{\partial t} . \tag{5.27}$$

The Poisson equation

$$\frac{\partial E}{\partial x_L} = (N - n_L) \tag{5.28}$$

and the definition

$$\frac{dx_L}{dt} = v_L = \frac{p_L}{\gamma} \tag{5.29}$$

lead together with (5.26) to

$$\frac{dE_L}{dt} = Nv_L . \tag{5.30}$$

We treat the case of homogeneous ion density N first. After integration we have

$$E_L = Nx_L + E_0(x_0) - Nx_0 , \tag{5.31}$$

where we have defined $E_0(x_0) = E_L(x_0, t = 0)$. Equation (5.31) is a basic relation which we shall use below.

5.3. Wave-breaking calculations in Lagrangian coordinates

The (longitudinal) momentum balance reads

$$\frac{dp_L}{dt} = -E_L = -Nx_L - E_0(x_0) + Nx_0 \ . \tag{5.32}$$

Together with Eq. (5.29) this forms the basic dynamical equations for constant ion density. Here,

$$\gamma = \sqrt{1 + p_L^2} \ . \tag{5.33}$$

When the ion density distribution is not homogeneous (but the ion dynamics is still neglected) we have $N_L = N_L(x_0, t) = N(x_L)$. The generalization of Eq. (5.30) is

$$\frac{dE_L}{dt} = N_L \frac{dx_L}{dt} \ . \tag{5.34}$$

Introducing the function $Y_i(\xi)$ such that

$$N(x_L) = \left.\frac{dY_i}{d\xi}\right|_{\xi=x_L} \tag{5.35}$$

we find

$$E_L = [Y_i(x_L) - Y_i(x_0)] + E_0(x_0) \ , \tag{5.36}$$

which replaces the previous equation (5.31).

Now equation (5.32) can be generalized to

$$\frac{dp_L}{dt} = -[Y_i(x_L) - Y_i(x_0)] - E_0(x_0) \ , \tag{5.37}$$

which, together with (5.29) forms the basic system of equations for inhomogeneous plasmas.

5. Relativistic wave-breaking in cold plasma

5.3.1 Wave-breaking analysis

We now exemplify wave-breaking for N=const; the conclusions and calculations are the same for inhomogeneous situations $N(x)$.

Using Poisson's equation we have

$$\frac{\partial E_L}{\partial x_0} = \frac{\partial E}{\partial x_L}\frac{\partial x_L}{\partial x_0} = (N - n_L)\frac{\partial x_L}{\partial x_0} \ . \tag{5.38}$$

On the other hand

$$\frac{\partial E_L}{\partial x_0} = \frac{\partial}{\partial x_0}(Nx_L + E_0(x_0) - Nx_0) \ . \tag{5.39}$$

A short calculation leads to

$$\frac{\partial E_L}{\partial x_0} = \left(\frac{\partial x_L}{\partial x_0}N - n_0\right) \tag{5.40}$$

where we have used

$$\frac{\partial E_0}{\partial x_0} = (N - n_0) \tag{5.41}$$

with $n_0 = n_L(x_0, t = 0)$. Comparing the right-hand-sides of (5.38) and (5.40) we find

$$n_L = \frac{n_0}{\frac{\partial x_L}{\partial x_0}} \ . \tag{5.42}$$

The condition

$$\frac{\partial x_L}{\partial x_0} = 0 \tag{5.43}$$

defines wave-breaking. Relation (5.42) shows that infinite density corresponds to wave-breaking.

For the case N=const, we insert into (5.32) the definition

$$y_L := Nx_L + E_0 - x_0 N \ . \tag{5.44}$$

5.3. Wave-breaking calculations in Lagrangian coordinates

We then can write the basic dynamical equations as

$$\frac{dy_L}{dt} = N\frac{p_L}{\gamma}, \tag{5.45}$$

$$\frac{dp_L}{dt} = -y_L. \tag{5.46}$$

Because of the appearance of γ this is a nonlinear oscillator (let us assume $N = 1$)

$$\frac{d^2 p_L}{dt^2} = -\frac{p_L}{\sqrt{1+p_L^2}}. \tag{5.47}$$

Instead of ω_{pe0} in general an amplitude-dependent frequency appears.

The amplitude-dependent frequency can be determined by perturbation analysis. First we expand the square-root for small amplitudes, to obtain for the oscillator (with linear frequency $\omega = 1$)

$$\frac{d^2 p}{dt^2} + \omega^2 p = -(1-\omega^2)p - \frac{1}{2}p^3 \tag{5.48}$$

Multiplication of both sides with $\cos(\omega t)$ and integration over t from $-\pi/\omega$ to $+\pi/\omega$, and approximating $p(t) \approx \tilde{A}\cos(\omega t)$ within the integrals, leads to the approximate result for the frequency at small amplitudes

$$\omega^2 \approx 1 - \frac{3}{8}\tilde{A}^2 \tag{5.49}$$

This is the result shown in Eq. (5.70). It is depicted in Fig. 5.3 by the blue curve.

We can even obtain a better approximation, being valid for larger amplitudes, when not expanding the square-root. By performing the same steps as before, we arrive at

$$\omega^2 \approx \frac{1}{\pi}\int_{-\pi}^{\pi} d\tau \frac{\cos^2(\tau)}{\sqrt{1+\tilde{A}^2\cos^2(\tau)}} \tag{5.50}$$

5. Relativistic wave-breaking in cold plasma

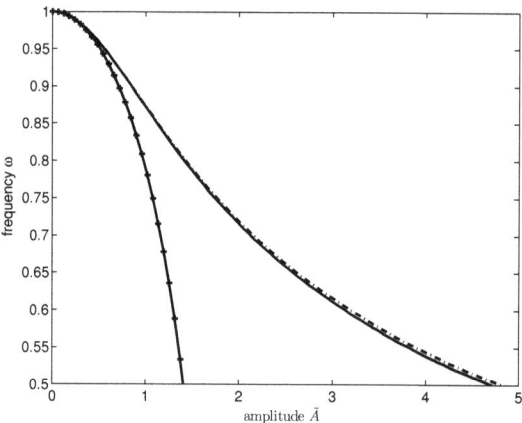

Figure 5.3: Amplitude dependence of the frequency of the nonlinear oscillator (5.47). The exact result (solid line) is compared with the small amplitude approximation (5.49) (crossed line) and (5.50) (dash–dotted curve).

This result is also shown in Fig. 5.3 by the red curve. When compared to the exact numerical result, we recognize excellent agreement up to large amplitudes.

For inhomogeneous $N = N(x)$ an approximate calculation is possible, e.g. when $|x_L - x_0| \ll 1$. Then we start from the equations

$$\frac{dp_L}{dt} = -N(x_0)[x_L - x_0] - E_0(x_0) \equiv -\tilde{y}_L , \qquad (5.51)$$

$$\frac{d\tilde{y}_L}{dt} = \frac{N(x_0)}{\gamma} p_L . \qquad (5.52)$$

Now, the frequency is not only amplitude-, but also space-dependent. The solutions for $\tilde{y}_L$ and $x_L - x_0$ are straightforward to obtain.

5.3. Wave-breaking calculations in Lagrangian coordinates

Non-relativistic limit $\gamma = 1$, N=const

Let us assume $N = 1$, i.e. the ion density is constant and equal to the electron background density (this is not a necessary restriction of the model, although it simplifies the expressions). If we start with an harmonic field at $t = 0$

$$E_0 = A(x_0)\sin(kx_0 + \varphi) \tag{5.53}$$

and momentum

$$p_0 = -A(x_0)\cos(kx_0 + \varphi), \tag{5.54}$$

where φ is an arbitrary phase, we can write the wave-solution as

$$E_L = A(x_0)\sin(kx_0 - \omega_{pe0}t + \varphi). \tag{5.55}$$

The reason is that in the present case Eqs. (5.45) and (5.46) simplify to

$$\frac{d^2 E_L}{dt^2} = -\omega_{pe0}^2 E_L. \tag{5.56}$$

Then the basic relation (5.31) leads to

$$x_L = x_0 + A(x_0)\left[\sin(kx_0 - \omega_{pe0}t + \varphi) - \sin(kx_0 + \varphi)\right], \tag{5.57}$$

and we have for constant amplitudes $A(x_0) = A$

$$\frac{\partial x_L}{\partial x_0} = 1 + kA\left[\cos(kx_0 - \omega_{pe0}t + \varphi) - \cos(kx_0 + \varphi)\right]. \tag{5.58}$$

5. Relativistic wave-breaking in cold plasma

A necessary condition for wave-breaking is

$$kA \geq \frac{1}{2}, \tag{5.59}$$

which corresponds to the previous finding [19] (note the factor 2). Wave-breaking occurs when the peak fluid velocity change is larger than the phase velocity, i.e.

$$\Delta v_L|_{max} = 2A > \frac{1}{k}. \tag{5.60}$$

Dawson [19] has obtained the result $A \geq 1/k$ for specific initial conditions. For plasma waves with $k_p = 2\pi/\lambda_p$ this condition is equivalent to the result of $E_0 = cm_e\omega_p/e$ (in dimensional quantities) as maximum electrical field E_0 that a nonrelativistic cold plasma can sustain.

Let us make the following remarks on the famous result (5.59). First, we may use (5.59) for determining the wave-breaking onset in dependence of, e.g., the amplitude A. Introducing $T = \omega_{pe0}^{-1}$, $V = \Delta v_L|_{max}$, and $L = k^{-1}$ we find from

$$VT \approx L \tag{5.61}$$

the threshold amplitude for fixed k. When approximating *for wake-fields* $\frac{\omega_{pe0}}{k} \approx 1$ and using the maximum momentum amplitude A, the criterion (5.59) reads

$$A \geq \frac{1}{2}. \tag{5.62}$$

Thus, according to (5.59) small amplitudes do not lead to wake-field breaking. On the other hand, (5.62) for wake-fields requires a relativistic treatment.

In the homogeneous non-relativistic limit (not necessarily wake-fields) we have compared the threshold prediction (5.59) for wave-breaking in finite time with numerical

5.3. Wave-breaking calculations in Lagrangian coordinates

simulations and found excellent agreement.

Relativistic case

In the fully relativistic description we obtain instead of (5.57) a nonlinear oscillator solution which we abbreviate in the form

$$x_L = x_0 + A\sin(kx_0)\left[F(\omega t, x_0) - 1\right] , \tag{5.63}$$

where ω is the nonlinear frequency which in the weakly relativistic limit is $\omega \approx \omega_{pe0}$. The function F is still finite and 2π-periodic, i.e. $F(y \equiv \omega t + 2\pi, x_0) = F(y \equiv \omega t, x_0)$. When we start with $E_0 = A\sin(kx_0)$, differentiation now leads to

$$\frac{\partial x_L}{\partial x_0} = 1 + \left\{\frac{\partial E_0}{\partial x_0}\left[F(y \equiv \omega t, x_0) - 1\right] + E_0 \frac{\partial F(y \equiv \omega t, x_0)}{\partial y}\frac{\partial \omega}{\partial x_0} t + E_0 \frac{\partial F(y \equiv \omega t, x_0)}{\partial x_0}\right\} \tag{5.64}$$

The right-hand-side contains, in addition to the Dawson criterion (5.59), a second source for wave-breaking. In any case, for $t \to \infty$ the term being linear in t will dominate. For a given

$$\frac{\partial \omega}{\partial x_0} \lessgtr 0 \tag{5.65}$$

we can always find a series of intervals such that

$$E_0 \frac{\partial F(y \equiv \omega t, x_0)}{\partial y}\frac{\partial \omega}{\partial x_0} < 0 \tag{5.66}$$

and wave-breaking can occur without threshold. This modifies the statement [86] that "for both non-relativistic and relativistic plasmas, this overtaking happens when the peak fluid velocity equals the phase speed". The breaking considered here may not occur on the first oscillation, but quite later, i.e. after many electron plasma periods. In general, (5.66) allows to estimate the time for breaking. We would like to emphasize

5. Relativistic wave-breaking in cold plasma

that now the time for wave-breaking follows from

$$VT \approx L, \tag{5.67}$$

[see (5.61) which was used there for the amplitude threshold] where, however, now the inhomogeneity length

$$L \approx \left| \left[\frac{\partial \ln \omega}{\partial x_0} \right]^{-1} \right| \tag{5.68}$$

should be used. Quantitative predictions require the detailed knowledge of the nonlinear oscillation.

When does $\frac{\partial \omega}{\partial x_0} \lessgtr 0$ occur?

Obviously, the frequency becomes x_0-dependent when the plasma is inhomogeneous. Similarly, a nonlinearity can also introduce a space-dependent frequency. We demonstrate this analytically when, e.g., solving the (expanded) equation

$$\frac{d^2 p_L}{dt^2} = -\frac{\omega_{pe0}^2}{\gamma} p_L \approx -\omega_{pe0}^2 \left(1 - \frac{1}{2} p_L^2\right) p_L \tag{5.69}$$

for small amplitudes. For an initial momentum $p_0 \sim \tilde{A}$, we have approximately

$$\omega^2 \approx \omega_{pe0}^2 \left(1 - \frac{3}{8} \tilde{A}^2\right). \tag{5.70}$$

For example, with $\tilde{A} = \tilde{A}(x_0) = A(x_0) \cos(kx_0)$, we can determine the explicit space-dependence of the frequency.

Applying this to the wake-field being excited by a pulsating laser beam, we recognize that the initially excited wake-field has a space-dependent frequency. Thus the wake-field will break (sooner or later, depending on the strength of the exited field) even if (5.59) is not satisfied. For $\gamma \to 1$ the time for breaking tends to infinity. Using

5.3. Wave-breaking calculations in Lagrangian coordinates

wake-field relations, we can find the scaling

$$T \sim E_0^{-3} \tag{5.71}$$

in the small amplitude limit. Here E_0 is the maximum amplitude of the electrostatic wave. The result (5.59) can be understood as the (non-relativistic) prediction of breaking of (general) electrostatic oscillations in finite time. When we choose relativistic wake-fields, our numerical simulations always show wave-breaking in accordance with the analytical predictions.

Effect of finite field width

Dawson [19] considered a non-relativistic field of infinite length. To study the effect of finite field length, we start with an electric field at $t = 0$ in the form

$$E_L = A_\infty e^{-x_0^2/\sigma^2} \cos(kx_0) \sin(\omega_{pe0} t) \,. \tag{5.72}$$

The basic relation (5.31) leads to

$$x_L = x_0 + A_\infty \cos(kx_0) \sin(\omega_{pe0} t) e^{-x_0^2/\sigma^2} \,, \tag{5.73}$$

which after differentiation gives

$$\frac{\partial x_L}{\partial x_0} = 1 - A_\infty \left(k \sin(kx_0) + 2\frac{x_0}{\sigma^2} \cos(kx_0) \right) \sin(\omega_{pe0} t) e^{-x_0^2/\sigma^2} \,. \tag{5.74}$$

Therefore we expect breaking if

$$A_\infty \left(k \sin(kx_0) + 2\frac{x_0}{\sigma^2} \cos(kx_0) \right) e^{-x_0^2/\sigma^2} \geq 1 \,. \tag{5.75}$$

5. Relativistic wave-breaking in cold plasma

For $\sigma \to \infty$ we recover Dawson's criterion [19] for breaking, namely $A_\infty k \geq 1$. In the limit $\sigma \to 0$ the field is expected to break even for arbitrarily small field amplitudes. Between these two limits it is most convenient to evaluate the inequality (5.75) numerically. We define

$$\Phi \equiv \Phi(\sigma; k) = \frac{1}{\max_{x_0} \tilde{f}}, \quad \text{with} \quad \tilde{f} \equiv \tilde{f}(x_0; k, \sigma) = \left(\sin(kx_0) + 2\frac{x_0}{k\sigma^2}\cos(kx_0)\right) e^{-x_0^2/\sigma^2}. \tag{5.76}$$

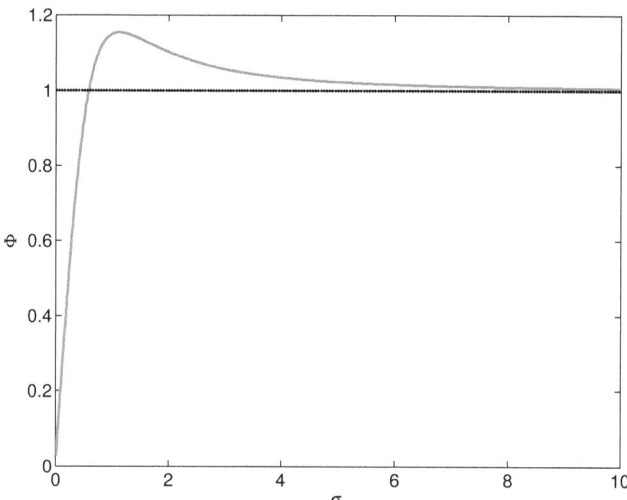

Figure 5.4: Numerical evaluation of the inequality (5.75) for $k = 2$. Fields with $A_\infty k \geq \phi$ are supposed to break.

In Fig. 5.4 the evaluation is shown for $k = 2$. The breaking criterion reads

$$A_\infty k \geq \Phi, \tag{5.77}$$

5.3. Wave-breaking calculations in Lagrangian coordinates

where, for simplicity, we treat k as a parameter. For very broad fields, i.e. large σ, the criterion resembles Dawson's result [19]. If σ gets smaller, some fields with amplitudes A_∞ fulfilling the breaking condition for infinite length, will break no more in the non-relativistic limit. For very small σ, even in the non-relativistic case, fields with very small amplitudes may suffer from breaking.

5.3.2 Numerical results for wave-breaking

We have checked most of the analytical predictions by numerical simulations. In the following, we present two typical examples.

Wave-breaking with fixed ion background

First, within a fully relativistic treatment, we solved the wave-breaking problem (5.45) and (5.46), i.e.

$$\frac{dE_L}{dt} = \frac{p_L}{\gamma}, \tag{5.78}$$

$$\frac{dp_L}{dt} = -E_L \tag{5.79}$$

for the special inhomogeneous initial conditions

$$E_L(x_0, t=0) = A_\infty e^{-x_0^2/\sigma^2} \cos(kx_0), \quad p_L(x_0, t=0) = 0. \tag{5.80}$$

From Eqs. (5.78) and (5.79) we obtain $E_L(x_0,t)$ and $p_L(x_0,t)$. When using (5.31) and (5.42), we can determine $n_e \equiv n(x_L)$ and $E(x_L)$. The *nonrelativistic* theory restricts breaking to $A_\infty k > \frac{1}{2}$ (for $\sigma \to \infty$). Choosing $k = 0.8$, according to that prediction breaking should only occur for $A_\infty \geq 0.625$. The simulation of the nonlinear oscillator equation already shows breaking for smaller amplitudes, e.g. for $A_\infty = 0.5$ as can be

5. Relativistic wave-breaking in cold plasma

seen in Fig. 5.5. Breaking below the threshold (5.59) supports our conclusions. The breaking time $t \approx 55$ occurs within the validity of the electron fluid model. The reason is that we are still close to the threshold (5.59).

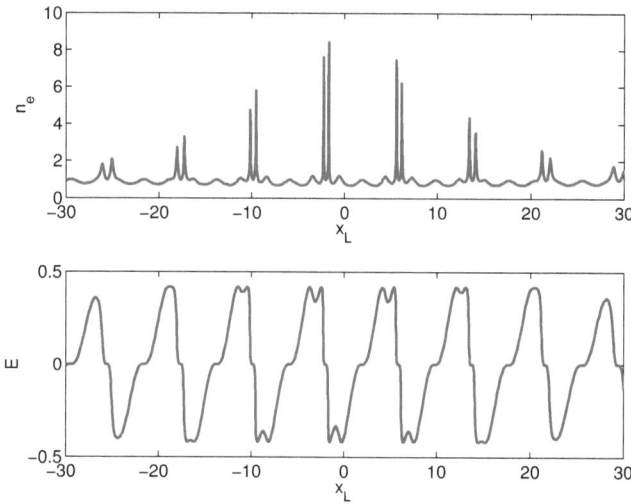

Figure 5.5: Electron density and electric field short before breaking at $t = 55$ when solving Eqs. (5.78) and (5.79). The initial electrical field is given by (5.80). The normalized parameters are $A_\infty = 0.5$, $k = 0.8$, and $\sigma = 50$.

Wave-breaking including ion dynamics

Next, we re-consider the wave-breaking of wake-fields which so far was investigated only within a purely electron fluid model. We have checked whether the ion (index i) dynamics will change the main conclusions. When ions are included, instead of Eqs. (5.20)–(5.23) the system of equations (4.3)–(4.6) can be used. A typical numerical

5.3. Wave-breaking calculations in Lagrangian coordinates

result is shown in Fig. 5.6.

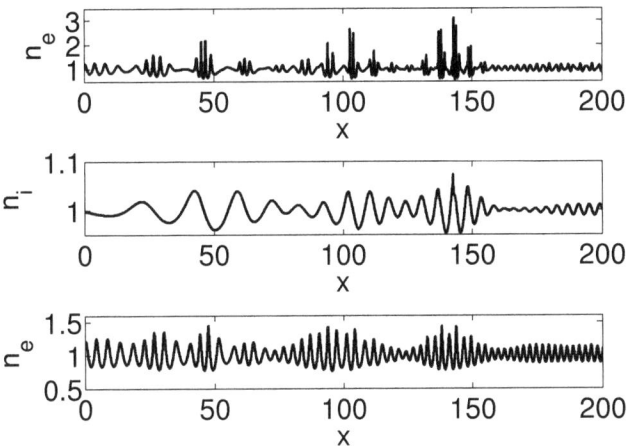

Figure 5.6: Same as Fig. 5.2, but now including the ions. The simulations are based on Eqs. (4.3) - (4.6). Electron and ion density of a wake-field are shown at $t = 1050$ at top and in the middle, respectively, shortly before breaking. The bottom graph shows for comparison the electron density at the same time, but now from Fig. 5.2 when ion motion is neglected.

All simulations show faster breaking when ion motion is included. Thus, the breaking criterion based on electron dynamics gives a proper upper limit for breaking.

Wave-breaking in warm plasmas

In warm plasma the equation for the parallel electron plasma momentum p reads

$$\frac{\partial}{\partial t}p = -\nabla\phi - \nabla\gamma - \frac{1}{n}\nabla P, \qquad (5.81)$$

5. Relativistic wave-breaking in cold plasma

where P is the scalar pressure in an isotropic plasma. To close the system, an equation of state has to be introduced. This could be for example $P = nT$ for an *isothermal* or $P = (n/\gamma)^3 T$ [80] for an *adiabatic* plasma. Equation (5.81) is obtained from warm fluid theories and therefore is only valid for systems in which the particle thermal velocities are not of the order of the speed of light. Very high temperatures are therefore not treated correctly by this description, hence a fully relativistic would be necessary to treat such problems.

As shown in the last sections, very steep gradients are formed in the density when wave-breaking occurs in cold plasma. The scalar pressure term is proportional to ∇n in the *isothermal* plasma model, hence it can become very large and strongly influence the behavior near wave-breaking. In case of an *adiabatic* plasma model, the pressure should be even more beneficial, since it is proportional to $\nabla n^3 = 3n^2 \nabla n$.

In both plasma models the term for the pressure is proportional to $1/n \nabla n = 1/n \frac{\partial}{\partial x} n$, expressing this in in Lagrangian coordinates leads to

$$\begin{aligned}\frac{1}{n}\nabla P &\sim \frac{1}{n_L}\frac{\partial n_L}{\partial x_L} = \frac{1}{n_L}\frac{\partial n_L}{\partial x_0}\frac{\partial x_0}{\partial x_L} \\ &= \frac{1}{n_L}\frac{\partial n_L}{\partial x_0}\frac{1}{\frac{\partial x_L}{\partial x_0}} \\ &= \frac{1}{n_0}\frac{\partial n_L}{\partial x_0},\end{aligned} \quad (5.82)$$

where Eq. (5.42) has been used.

For a fixed ion background the oscillators which make up the wake-field in the Lagrangian description are governed by the equations

$$\frac{dE_L}{dt} = \frac{p_L}{\gamma}, \quad (5.83)$$

$$\frac{dp_L}{dt} = -E_L - \frac{1}{n_0}\frac{\partial n_L}{\partial x_0}. \quad (5.84)$$

5.3. Wave-breaking calculations in Lagrangian coordinates

The additional pressure term now spatially couples the oscillators that have been independent of each other so far.

The evaluation of the spatial derivative of n_L imposes severe restrictions on the spatial resolution of the grid, since $\partial n_L/\partial x_0$ will eventually become very large. In fact for reasonable temperatures, this term has to become so large that for any feasible discretization, it is always numerically unstable to use a spectral method to calculate the derivative. Only an adaptive method that switches between first-order derivatives in regions of steep gradient and a higher-order derivative in smooth regions could prevent oscillations that appear next to the strong gradient. Thus the question whether an additional pressure contribution may eventually stop the wave from breaking for arbitrarily (small) temperatures could not be answered numerically by the employed algorithms. Basically a two-scale problem arises here.

5. Relativistic wave-breaking in cold plasma

6 Two-dimensional dynamics of relativistic solitons

In chapter 4 the stability properties of a variety of 1D solitons were discussed. An important result was that solitons on the ion time-scale with a node number $p \geq 1$ are unstable with respect to small perturbations in a cold plasma. The typical time-scale for the growth of the instability is of the order of a few hundred $1/\omega_{pe}$. We found two different types of longitudinal instabilities whereof one was associated to a Raman like process [76].

To gain insight into the multidimensional evolution of solitons and possible instabilities, we will now extend the geometry to 2D. A fully nonlinear 3D description is not yet feasible for practical computing reasons, but a 2D model may already show important new features such as transversal instability, which may eventually result in pulse filamentation [77, 40, 9].

It is important to note that by 2D we mean that the variables depend only on two spatial coordinates. The fields themself are, of course, three-dimensional. The 2D models are two-dimensional in the mathematical sense, but physically they may even be three-dimensional if a certain symmetry, e.g. cylindrical symmetry, is assumed.

For sufficiently low laser intensities the system (4.3)-(4.6) can be reduced to a nonlinear Schrödinger equation (NLSE) for the laser pulse amplitude. The NLSE is the generic amplitude equation for many nonlinear physical systems, including water waves

6. Two-dimensional dynamics of relativistic solitons

in deep water, optical fiber transmission lines, quantum systems and laser propagation. For some of these systems longitudinally stable plane solitons are known and an analytical understanding of their transversal instability is available [46, 40]. From this analogy we may expect that the relativistic 1D solitons suffer from transversal instability, too.

In case of a transversal instability competing with a longitudinal instability, it is interesting to compare the growth rates of both instabilities in order to determine which one will prevail in the linear regime. As the nonlinear evolution does not necessarily resemble the linear regime we will have to confirm findings from the linear regime by full nonlinear 2D simulations.

6.1 Linearized 2D equations

Like in the 1D case we will perform a numerical linear stability analysis for the 2D system. We consider a laser pulse propagating along x and allow for transversal dependency of all quantities along y. We are free to chose an arbitrary transversal direction for the perturbation, so we choose one along a single transversal coordinate y. For practical reasons we introduce a co-moving frame by $\xi = x - Vt$, $\tau = t$. Then the 2D relativistic fluid-Maxwell equations are

$$\frac{\partial^2}{\partial \tau^2}\mathbf{A} - 2V\frac{\partial^2}{\partial \xi \partial \tau}\mathbf{A} - \left(1 - V^2\right)\frac{\partial^2}{\partial \xi^2}\mathbf{A} - \frac{\partial^2}{\partial y^2}\mathbf{A} = \mathsf{P}^{df}\mathbf{j}, \quad (6.1)$$

$$\left(\frac{\partial^2}{\partial \xi^2} + \frac{\partial^2}{\partial y^2}\right)\phi = n_e - n_i, \quad (6.2)$$

$$\frac{\partial}{\partial \tau}n_\alpha - V\frac{\partial}{\partial \xi}n_\alpha + \frac{\partial}{\partial \xi}j_x + \frac{\partial}{\partial y}j_y = 0, \quad (6.3)$$

6.1. Linearized 2D equations

$$\frac{\partial}{\partial \tau} \begin{pmatrix} M_{\alpha x} \\ M_{\alpha y} \\ M_{\alpha z} \end{pmatrix} - V \frac{\partial}{\partial \xi} \begin{pmatrix} M_{\alpha x} \\ M_{\alpha y} \\ M_{\alpha z} \end{pmatrix} - \frac{\epsilon_\alpha}{\gamma_\alpha} \begin{pmatrix} p_{\alpha y} (\partial_\xi M_{\alpha y} - \partial_y M_{\alpha x}) + p_{\alpha z} \partial_\xi M_{\alpha z} \\ p_{\alpha z} \partial_y M_{\alpha z} - p_{\alpha x} (\partial_\xi M_{\alpha y} - \partial_y M_{\alpha x}) \\ -p_{\alpha x} \partial_\xi M_{\alpha z} - p_{\alpha y} \partial_y M_{\alpha z} \end{pmatrix}$$

$$= \begin{pmatrix} \partial_\xi \\ \partial_y \\ 0 \end{pmatrix} \left(q_\alpha \phi - \frac{\gamma_\alpha}{\epsilon_\alpha} \right) , \qquad (6.4)$$

$$\mathbf{j} = \varepsilon_i \frac{n_i \mathbf{P}_i}{\gamma_i} - \frac{n_e \mathbf{P}_e}{\gamma_e} . \qquad (6.5)$$

$\mathsf{P}^{df} = 1 - \nabla (\nabla^2)^{-1} \nabla \cdot$ is a projector which gives the divergence-free part of a given vector field. In this notation $\nabla = (\partial_\xi, \partial_y, 0)^T$.

Linearization about the unperturbed solution and subsequent Fourier-transformation in y direction (i.e. $\partial_y \to k_y$, if we assume only a oscillation in y direction with wavenumber k_y) gives (indices according to Sec. 3.3)

$$\frac{\partial^2}{\partial \tau^2} \mathbf{A}_1 - 2V \frac{\partial^2}{\partial \xi \partial \tau} \mathbf{A}_1 - \left(1 - V^2\right) \frac{\partial^2}{\partial \xi^2} \mathbf{A}_1 + k_y^2 \mathbf{A}_1 = \mathsf{P}^{df} \left(\mathbf{j}_{i1} - \mathbf{j}_{e1} \right) , \qquad (6.6)$$

$$\left(\frac{\partial^2}{\partial \xi^2} - k_y^2 \right) \phi_1 = n_{e1} - n_{i1} , \qquad (6.7)$$

$$\frac{\partial}{\partial \tau} n_{\alpha 1} - V \frac{\partial}{\partial \xi} n_{\alpha 1} + \frac{\partial}{\partial \xi} j_{\alpha 1 x} + i k_y j_{\alpha 1 y} = 0 , \qquad (6.8)$$

6. Two-dimensional dynamics of relativistic solitons

$$\frac{\partial}{\partial t}\mathbf{M}_{\alpha 1} - V\frac{\partial}{\partial x}\mathbf{M}_{\alpha 1}$$
$$- \frac{\epsilon_\alpha}{\gamma_{\alpha 0}}\left[-\frac{1}{\gamma_{\alpha 0}}\gamma_{\alpha 1}\mathbf{p}_{\alpha 0} \times (\nabla \times \mathbf{M}_{\alpha 0}) + \mathbf{p}_{\alpha 1} \times (\nabla \times \mathbf{M}_{\alpha 0}) + \mathbf{p}_{\alpha 0} \times (\nabla \times \mathbf{M}_{\alpha 1})\right]$$
$$= \begin{pmatrix} \partial_\xi \\ \mathrm{i}k_y \\ 0 \end{pmatrix}\left(q_\alpha\phi_1 - \frac{1}{\epsilon_\alpha}\gamma_{\alpha 1}\right), \quad (6.9)$$

where

$$\mathbf{j}_{\alpha 1} = \frac{\epsilon_\alpha}{\gamma_{\alpha 0}}\left[n_{\alpha 1}\mathbf{p}_{\alpha 0} + n_{\alpha 0}\mathbf{p}_{\alpha 1} - \frac{1}{\gamma_{\alpha 0}}n_{\alpha 0}\gamma_{\alpha 1}\mathbf{p}_{\alpha 0}\right], \quad (6.10)$$

$$\gamma_{\alpha 1} = \epsilon_\alpha^2 \frac{\mathbf{p}_{\alpha 0} \cdot \mathbf{p}_{\alpha 1}}{\gamma_{\alpha 0}}. \quad (6.11)$$

Since the solitons which we were examining as unperturbed states all have a zero generalized transversal momentum, $\mathbf{M}_{\alpha 0\perp} = 0$, the entries of the vector products are

$$\mathbf{p}_{\alpha 0} \times (\nabla \times \mathbf{M}_{\alpha 1}) = \begin{pmatrix} p_{\alpha 0y}\left(\partial_\xi M_{\alpha 1y} - \mathrm{i}k_y M_{\alpha 1x}\right) + p_{\alpha 0z}\partial_\xi M_{\alpha 1z} \\ \mathrm{i}k_y p_{\alpha 0z} M_{\alpha 1z} - p_{\alpha 0x}\left(\partial_\xi M_{\alpha 1y} - \mathrm{i}k_y M_{\alpha 1x}\right) \\ -p_{\alpha 0x}\partial_\xi M_{\alpha 1z} - \mathrm{i}k_y p_{\alpha 0y} M_{\alpha 1z} \end{pmatrix}, \quad (6.12)$$

$$\mathbf{p}_{\alpha 0} \times (\nabla \times \mathbf{M}_{\alpha 0}) = \begin{pmatrix} -\mathrm{i}k_y p_{\alpha 0y} M_{\alpha 0x} \\ \mathrm{i}k_y p_{\alpha 0x} M_{\alpha 0x} \\ 0 \end{pmatrix}, \quad (6.13)$$

$$\mathbf{p}_{\alpha 1} \times (\nabla \times \mathbf{M}_{\alpha 0}) = \begin{pmatrix} -\mathrm{i}k_y p_{\alpha 1y} M_{\alpha 0x} \\ \mathrm{i}k_y p_{\alpha 1x} M_{\alpha 0x} \\ 0 \end{pmatrix}. \quad (6.14)$$

6.2. Transversal instability

For any fixed k_y we end up with a stability problem formulated in form of a system of one-dimensional partial differential equations. This reduction in dimensionality vastly decreases the computational effort and allows to determine the growth rate for transversal instabilities with different k_y in a reasonable time.

The algorithms to numerically solve the nonlinear and the linearized 2D equations are the same as described in Sec. 2.3.

6.2 Transversal instability

For the 1D set of equations (4.3) - (4.6) circular polarized soliton solutions (with and without included ion motion) are known [39, 26]. We studied the influence of transversal (to the propagation direction) perturbations on the Bulanov-Farina solitons, which were derived in Sec. 4.3.2.

The longitudinal stability properties have been examined for these solitons with a fixed ion background in Ref. [76] and for mobile ions in Section 4.3.2 and Ref. [54].

In Sec. 4.3.2 we analyzed the longitudinal stability properties of Bulanov-Farina solitons, depending on the number of nodes p. For solitons with $p = 0$ we did not find any physically relevant longitudinal instability, whereas for $p \geq 1$ we were able to find exponentially growing modes with notable growth rates. Depending on the velocity we were able to distinguish two types of longitudinal instabilities, purely growing ones and oscillatory growing ones.

Within the linearized 2D model we were looking for exponentially growing modes, growing proportional to $\exp(\Gamma t)$.

6. Two-dimensional dynamics of relativistic solitons

6.2.1 Transversal instability of $p = 0$ solitons

It was presented in Sec. 4.3.2 that $p = 0$ solitons show no longitudinal instability. From the simulation of the linearized system (6.6)-(6.9) we can derive transversely unstable modes for all velocities V that are possible.

To study the transversal stability of a soliton we performed simulations of Eq. (6.6) - (6.9) for different wave-numbers $k_\perp$ of the transversal perturbation. We got the structure for the most unstable mode for this particular wave-number $k_\perp$ and its growth rate Γ. In general, the growth rate is a function of the wave-number. By performing a series of simulations with increasing $k_\perp$, starting from $k_\perp = 0$, we were able to determine the dependence of Γ from $k_\perp$. Figure 6.1 depicts typical results for $p = 0$ solitons. Results are presented for two different velocities and three different ω_0 values for each velocity. In the limit of infinite wavelength for the perturbation, we expect the growth rate to be zero. Going to smaller wavelengths, increasing $k_\perp$, we first see a linear increase in the growth rate Γ. Eventually the profile of $\Gamma(k_\perp)$ deviates from linear characteristics and forms a global maximum. For even larger $k_\perp$ the growth rates become rapidly smaller, until the cut-off wave-number $k_{\perp c}$ is reached. For wave-numbers larger than $k_{\perp c}$ (hence wavelengths smaller than $\lambda_{\perp c} = 2\pi/k_{\perp c}$), no exponentially growing instabilities within the linear regime are found. However perturbations with wave-numbers above the cut-off can grow in the nonlinear regime, where energy may be transferred to these modes by a wave-vector cascade. Finding the exact value of the cut-off wave-number from simulations is difficult, because $\partial\Gamma/\partial k_\perp$ is very large in the proximity of this point. To us the exact position of the cut-off is of minor importance, our focus will be on the transversal wave-number for which we obtained the mode with the largest growth rate Γ. The mode with the largest growth rate will play the dominating role in the dynamics of a general perturbation in the linear regime. This will be demonstrated in Sec. 6.3.3. All Γ are real numbers

6.2. Transversal instability

for the transversal instability (which is different to the longitudinal case), hence the perturbation is purely growing in time and shows no additional oscillations over time.

This kind of profile for Γ varying with $k_\perp$ is known from other solitons, e.g. solitons of the 1D nonlinear Schrödinger equation [46]. In fact for very fast solitons the amplitude of the Bulanov-Farina solitons is very small, the ion response can be neglected and we can reduce the 1D equations (4.3)-(4.6) (for which the solitons are stationary solutions) to a 1D nonlinear Schrödinger equation. As stationary solutions we recover known bright soliton solutions for which an analytical description of the transversal instability is available [46].

For $p = 0$ we have the special circumstance that for every possible soliton velocity V there is a range in the frequency ω_0 for which the solitons exist. We investigated solitons with different frequencies at fixed velocity and found that the maximum growth rate increases as the frequency decreases, which can also be seen from Figure 6.1. In general the maximum growth rate of all perturbations is becoming larger the slower the solitons are. The value of $k_\perp$ for which this maximum growth rate is attained decreases as the velocity increases.

The maximum growth rates are quite significant, since a growth rate of about $\Gamma = 0.035$ (for a soliton with $V = 0.1$, $\omega_0 = 0.97$ and $k_\perp = 0.27$) would allow a 1% perturbation of this wavelength to grow to the order of one in a time of only $t \sim 135$.

Summarizing, the longitudinal stable $p = 0$ Bulanov-Farina solitons are unstable for transversal perturbations. The growth rate allows significant growth of the perturbation within physically relevant times, which are in the order of a few to several $1/\omega_{pi}$.

6. Two-dimensional dynamics of relativistic solitons

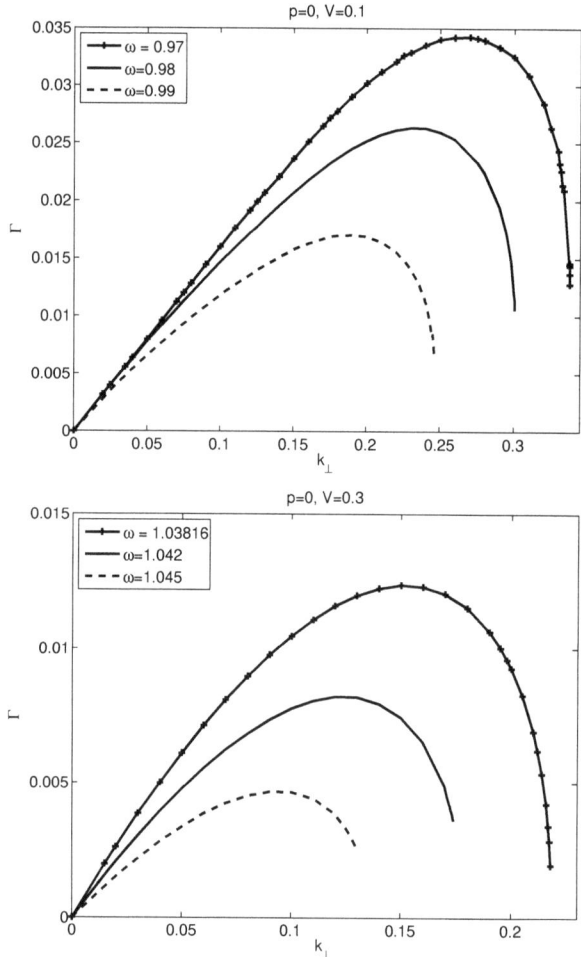

Figure 6.1: Linear growth rates Γ for two $p = 0$ solitons with velocities $V = 0.1$ (top) and $V = 0.3$ (bottom). For each velocity the growth rates for the transversal instability with wave numbers $k_\perp$ are shown for three solitons which differ in their frequencies ω_0.

6.2. Transversal instability

6.2.2 Linear transversal instability of p=1,2,... solitons

We have shown in Sec. 4.3.2 that Bulanov-Farina solitons with node number $p \geq 1$ in the vector potential suffer from longitudinal instability. For $p = 1$ solitons the growth rates for the 1D instability range from $\Gamma \approx 0.01$ to $\Gamma \approx 0.09$, depending on the soliton velocity. Perturbations of solitons with higher node numbers have growth rates with the same order of magnitude.

From simulations of Eq. (6.6) - (6.9) performed for $p = 1, 2, ...$ solitons as unperturbed solutions, we found that they show a transversal instability, too. To determine the transversal wave-number for which the growth rate Γ is maximal we used the same method as for the $p = 0$ solitons. All growth rates Γ have no imaginary part.

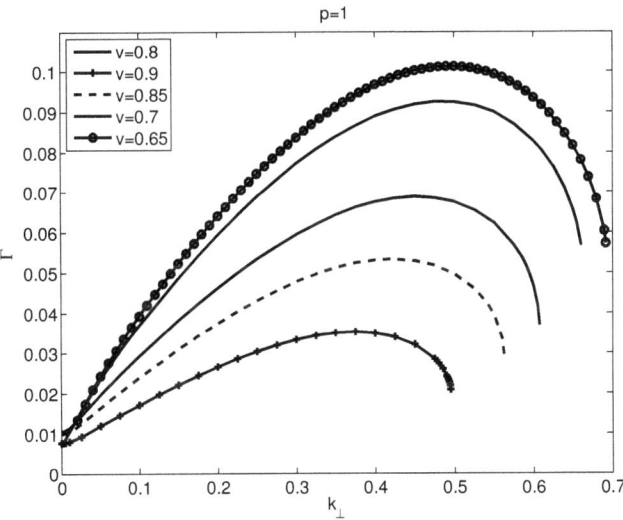

Figure 6.2: Linear growth rates Γ for $p = 1$ solitons with different velocities V versus the perturbation wave-number $k_\perp$.

6. Two-dimensional dynamics of relativistic solitons

The qualitative findings are the same as for the $p = 0$ case, see Fig. 6.2. The growth rates for the transversal instability are between $\Gamma \approx 0.01$ and $\Gamma \approx 0.15$ for $p = 1$ solitons, depending on the velocity V. The slower the soliton, the larger is the maximum growth rate Γ and the larger is $k_\perp$ for which this growth rate is attained. When we compare solitons with different p number but the same velocity, we find that the maximum growth rate increases with increasing node number p. Note that since these solitons are longitudinally unstable, we expect to recover the growth rate for the longitudinal instability in the limit of $k_\perp \to 0$. This behavior can be observed from Fig. 6.2.

In order to compare the importance of the transversal versus the longitudinal instability in the linear regime, we calculated the ratio of the growth rates. The maximum transversal growth rate is always larger than the longitudinal growth rate by a factor of at least 2 (up to a factor of 14). Figure 6.3 shows the ratio of the (transversal over longitudinal) growth rates in dependency of the soliton velocity V.

The peak in Figure 6.3 is due to a minimum in the longitudinal growth rate, which separates two regions of different unstable behaviors in the 1D case. For velocities smaller than $V \approx 0.7$ the 1D unstable mode shows frequency sidebands. On the other hand solitons faster than $V \approx 0.7$ the instability shows no frequency sidebands.

From the comparison of the linear growth rates for longitudinal and transversal instability we can state, that in the linear regime the transversal instability always dominates the dynamics. This however has not to hold for the nonlinear regime, where nonlinearities can influence this behavior.

6.3 Nonlinear simulations

To verify our findings from the linear simulation, we carried out nonlinear 2D simulations. As initial conditions a 1D soliton is used which is constant along y. A small

6.3. Nonlinear simulations

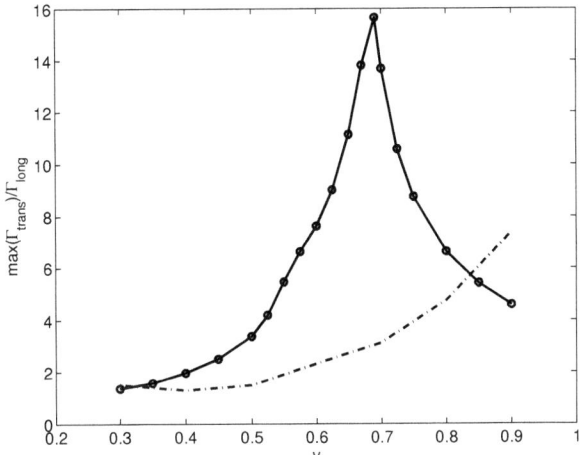

Figure 6.3: Ratio of the maximum transversal growth rate and the longitudinal growth rate for $p = 1$ (solid line) and $p = 2$ (dashed line) solitons with different velocities V. The peak for $p = 0$ is associated with a minimum in the longitudinal growth rate which separates solitons suffering from different types of longitudinal instability.

6. Two-dimensional dynamics of relativistic solitons

amount of a few percent of the most unstable mode (for a given $k_\perp$) is then added to this distribution. In the following sections we will demonstrate this for cases when the wave numbers $k_\perp$ has been chosen such that the growth rate has a maximum.

The comparison with the linear integrator leads to an independent verification of the linear result. As long as the perturbation is small, linear and nonlinear evolution have to be the same. All our linear results have been verified by this method and gave excellent agreement with the nonlinear results.

Eventually the perturbation will become large and the nonlinear evolution will set in. Because of their different stability properties in the purely longitudinal case, we distinguish in this section between nodeless solitons and solitons with nodes.

6.3.1 Nodeless solitons

We show in the following the nonlinear evolution of a perturbation for a $p = 0$ soliton with velocity $V = 0.1$. The norm of the perturbation added to the unperturbed soliton was 2% of the norm of the soliton. The initial evolution of the longitudinal perturbation is in excellent agreement with our linear prediction. When the nonlinear development prevails, a complete transversal filamentation of the perturbed soliton appears. The electrons start being expelled from the regions where the intensity $|A|^2$ is large and begin to bunch sideways of the pulse. The ion density forms peaks at the front and is depleted inside the filaments, see Fig. 6.4.

6.3.2 Solitons with nodes

Since solitons with $p = 1, 2...$ are transversely as well as longitudinally unstable, it is important to check the nonlinear evolution of these instabilities. This evolution of the two instabilities could be different from the linear one, which is described by the ratio of the linear growth rates from Sec. 6.2.2.

6.3. Nonlinear simulations

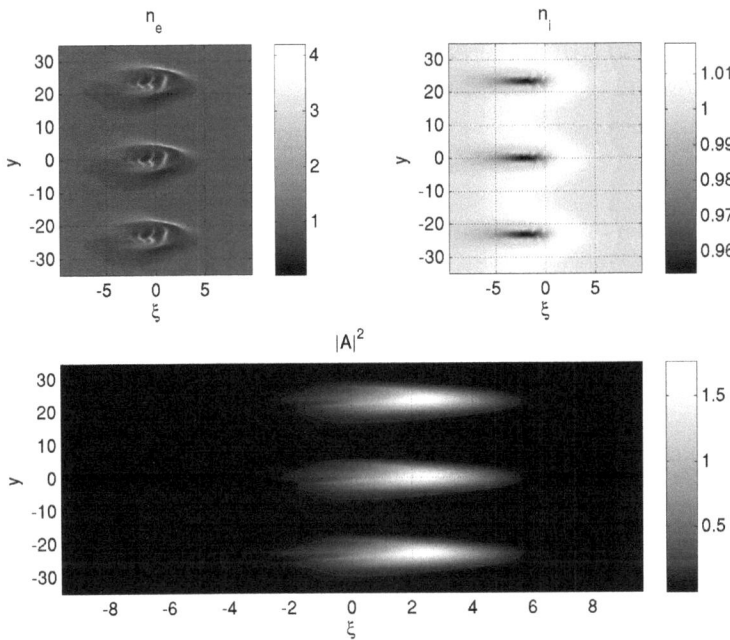

Figure 6.4: Densities n_e, n_i and intensity $|A|^2$ of a transversal perturbed $p = 0$ soliton with $V = 0.1$ at time $t = 105$. The perturbation is 2% of the transversal most unstable mode which has a wave number $k_\perp = 0.27$ and a growth rate $\gamma = 0.034$.

6. Two-dimensional dynamics of relativistic solitons

The simulations show that the transversal perturbation always prevails over the longitudinal one in the nonlinear regime. We demonstrate this for a $p = 1$ soliton with $V = 0.8$. The initial perturbations were constructed in such a way that they contain each of the most unstable modes for every direction, as determined by our linear simulations. The norm of the perturbations in each direction were 1% of the norm of the soliton. The longitudinal growth rate is $\Gamma_l = 0.010393$ and the transversal $\Gamma_t = 0.06890$. From Figure 6.5 can be seen that after a time of 56 the pulse already shows clear transversal filamentation, but almost no longitudinal deformation.

6.3.3 Transversal instability from noise

So far we always assumed knowledge of the most unstable modes from the linearized systems. To demonstrate the development of the transversal instability for systems without correct initial excitation of these modes, we simulate a soliton with random noise added to the initial vector potential $\mathbf{A}$. The noise $\mathbf{A}_{noise}$ is Gaussian distributed about a mean value. Besides fulfilling the Coulomb gauge $\nabla \cdot \mathbf{A}_{noise} = 0$ the amplitudes of the Fourier modes contained are such that the average magnitude of the corresponding magnetic field $\mathbf{B}_{noise}$ is the same for every wave number $\mathbf{k}_\perp = (k_\xi, k_y)$. Thus there is no preferred direction within the noise, and all Fourier modes of $\mathbf{B}_{noise}$ are equally strong. The energy of the added noise is about one permill of that of the total energy of the soliton. The highest frequency contained in the initial noise is in the order of a few percent of the highest frequency the computational grid can resolve. Initial excitation of larger frequencies would lead to strong aliasing errors from the numerical integration.

The simulations of such perturbed solitons show that within a time of the order of a few tens of $1/\omega_{pe}$ a transversal perturbation with a distinct $k_\perp$ dominates. The value of the transversal wave number $k_\perp$ corresponds to the maximum transversal

6.3. Nonlinear simulations

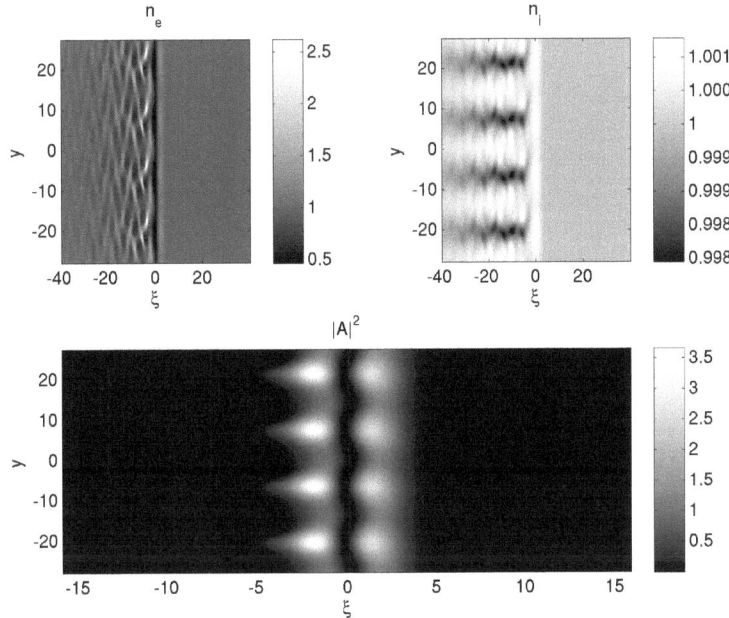

Figure 6.5: Densities n_e, n_i and intensity $|\mathbf{A}|^2$ of a perturbed $p = 1$ soliton with $V = 0.8$ at time $t = 56$. The perturbation consists of a longitudinal and a transversal mode. Each mode is the most unstable for its direction and the amount was 1% of the soliton for each. The dominance of the transversal instability can be seen.

6. Two-dimensional dynamics of relativistic solitons

growth rate that has been determined from the linear analysis. After the transversal instability has evolved for some time, wave-breaking sets in. This breaking is not due to a longitudinal instability, but is part of the transversal dynamics. Longitudinal instabilities can be excluded as source of this early breaking because of the different time scale on which the breaking sets in. The longitudinal growth rates are to small to allow noteworthy growing of these modes within the time considered here.

6.3. Nonlinear simulations

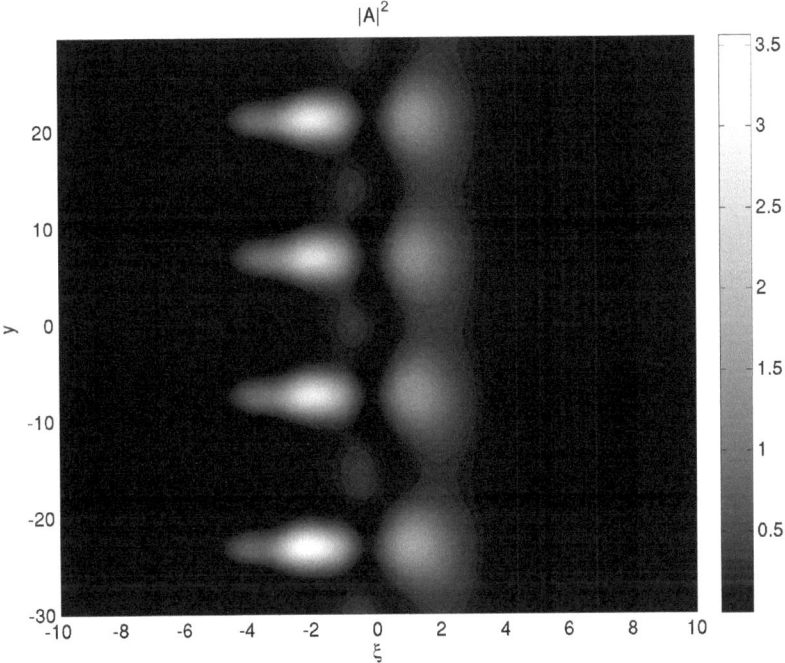

Figure 6.6: Intensity $|\mathbf{A}|^2$ of a transversal instability grown from noise for a $p = 1$ soliton with $V = 0.9$ at time $t = 290$. The energy of the noise was less than 1 permille of the energy of the soliton. The highest frequency of the initial noise was 5% of the maximum frequency the computational grid could resolve.

6. Two-dimensional dynamics of relativistic solitons

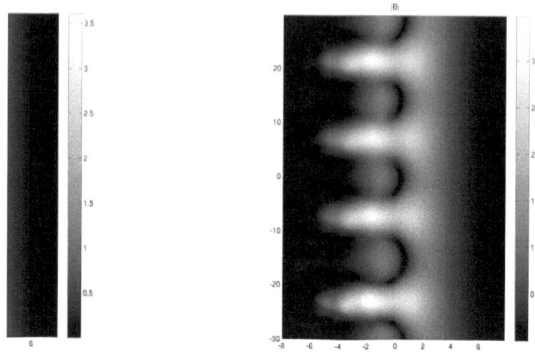

Figure 6.7: Magnitude of electric field E and magnetic field B at $t = 290$ for the instability shown in Fig. 6.6.

6.4 Field structure

In this section we will discuss the field structure of the perturbation and examine the nonlinear evolution of these structures.

Let us first recall the fields of the unperturbed soliton. The direction of propagation is x, the soliton is circular polarized and has vector potential components along y and z direction. The magnetic field $\mathbf{B} = \nabla \times \mathbf{A} = (0, B_y, B_z)$ of the soliton has only components perpendicular to the propagation direction. The electric field consists of an electrostatic part and an electromagnetic part. The electrostatic field is purely in x direction, whereas the electromagnetic contribution is solely in transversal direction. The magnetic and the electromagnetic fields oscillate with frequency ω_0.

The fields of the soliton depend only on one spatial coordinate, this is why we call it 1D soliton. The fields of the perturbation on the other hand depend on two spatial coordinates (but are of course three-dimensional themself). In two-dimensional

6.4. Field structure

geometry two kinds of polarization can be distinguished. Our simulation domain covers the x, y plane, which we call plane of incidence. The component of the electric field perpendicular to this plane is called s-polarized and the field components within this plane are called p-polarized (note that this p is a different letter than p, which we use to quantify the node number of a Farina-Bulanov soliton). By Maxwell's equation $\partial \mathbf{B}/\partial t = -\nabla \times \mathbf{E}$ we find the according magnetic field components. In Figure 6.8 this is shown schematically. There the fields B_x, B_y are combined into B_ϕ and E_x, E_y into E_ϕ. In the three-dimensional case the analog to the s-polarized (p-polarized) part is a TE-mode (TM-mode).

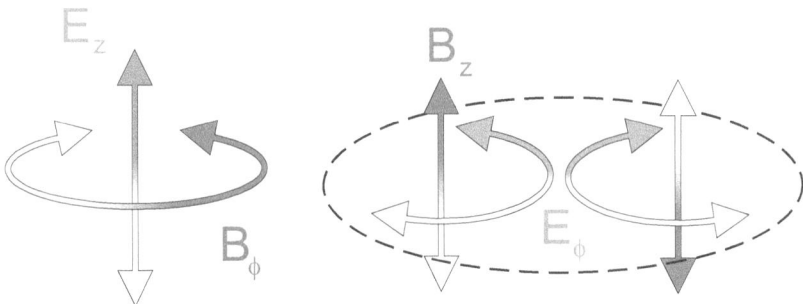

Figure 6.8: Schematic electromagnetic field configuration for a s-polarized (left) and a p-polarized (right) field. We assume the x, y plane as plane of incidence. The propagation direction is x.

We decomposed the fields of transversal perturbations into s- and p-polarized parts to study the structure of the perturbations. The distribution of the electromagnetic energy between these two parts is of interest to us.

The energy W in these fields is proportional to

$$W \sim \int \left(\mathbf{E}^2 + \mathbf{B}^2 \right) \mathrm{d}x \mathrm{d}y, \tag{6.15}$$

6. Two-dimensional dynamics of relativistic solitons

where we inserted the according field components for the different polarizations.

6.4.1 Fields of the linear perturbation

The two-dimensional perturbation has non-vanishing magnetic field components $\mathbf{B}_{pert} = (B_x, B_y, B_z)$. Due to an inhomogeneous distribution of electron and ion density the perturbation has an electrostatic field $\mathbf{E}_{es} = -\nabla\phi$, which has components along x and y direction. We define $\mathbf{E}_{em} = -\partial \mathbf{A}/\partial t$, which is the electric field of the laser. The fields $\mathbf{E}_{es}$ and $\mathbf{E}_{em}$ of the perturbation oscillate with the frequency ω_0 of the soliton.

The field $\mathbf{E}_{es}$ is not constant, but shows spatial variations over one soliton period, because of fluctuations in the electron density, see Fig. 6.9. The ion density shows no spatial fluctuation due to their much larger mass.

We split the electric field $\mathbf{E}_{em}$ and the magnetic field $\mathbf{B} = \nabla \times \mathbf{A}$ of the perturbation into p- and s-polarized parts.

Let us first discuss the s-polarized part. It consists of the fields B_x, B_y and E_z. In the x,y plane the components B_x and B_y form vortices, see Fig. 6.10 left side. These magnetic vortices are connected to an electric field by the Maxwell equation $\nabla \times \mathbf{B} = \partial \mathbf{E}_{em}/\partial t$. The electromagnetic field $E_{em,z}$ is depicted in the right side of Fig. 6.10. The field $E_{em,z}$ oscillates with the frequency ω_0, hence the curl direction oscillates with the same frequency. The phase difference of the oscillations in the electric and the magnetic field components is exactly $\pi/2$.

The field components $E_{em,x}, E_{em,y}$ and B_z constitute the p-polarized part of the perturbation. The magnetic field B_z has alternating signs in y direction which oscillate in time with ω_0. This oscillation is coupled to vortices in the electromagnetic field components $E_{em,x}$ and $E_{em,y}$ via $\nabla \times \mathbf{E} = \partial \mathbf{B}/\partial t$, see Fig. 6.11. As for the s-polarized part there is a phase difference of $\pi/2$ between the electric field and the magnetic field components.

6.4. Field structure

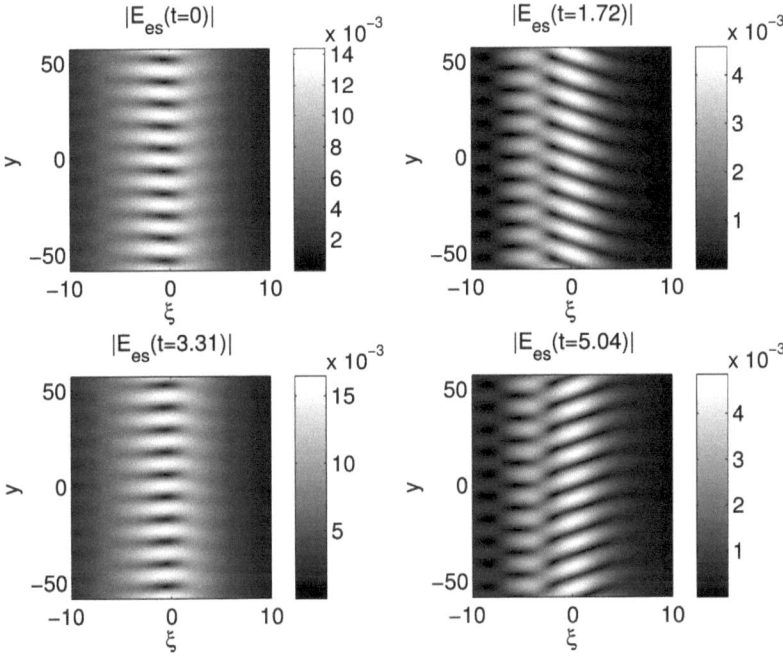

Figure 6.9: Contour plot of the magnitude of the electrostatic field $\mathbf{E}_{es}$ for a perturbation of a soliton with $p = 0$ and $V = 0.1$ for different times $t = 0$, $t = 1.72$, $t = 3.31$ and $t = 5.04$, respectively. The period of the oscillation is equal to the period of the soliton, which is $T = 6.57$. The wave-number of the perturbation is $k_\perp = 0.27$.

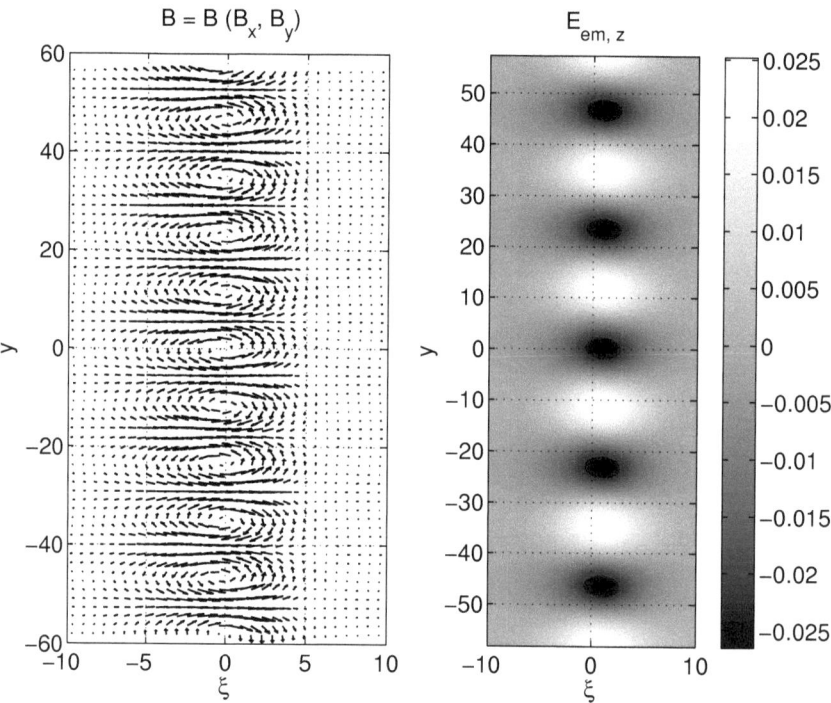

Figure 6.10: s-polarized field components of a perturbation of a soliton with $p = 0$ and $V = 0.1$. The wave-number of the perturbation is $k_\perp = 0.27$. Left: Field line plot of the B_ξ and B_y components of the magnetic field. Right: Contour plot of the z component of the electromagnetic part of the electric field.

6.4. Field structure

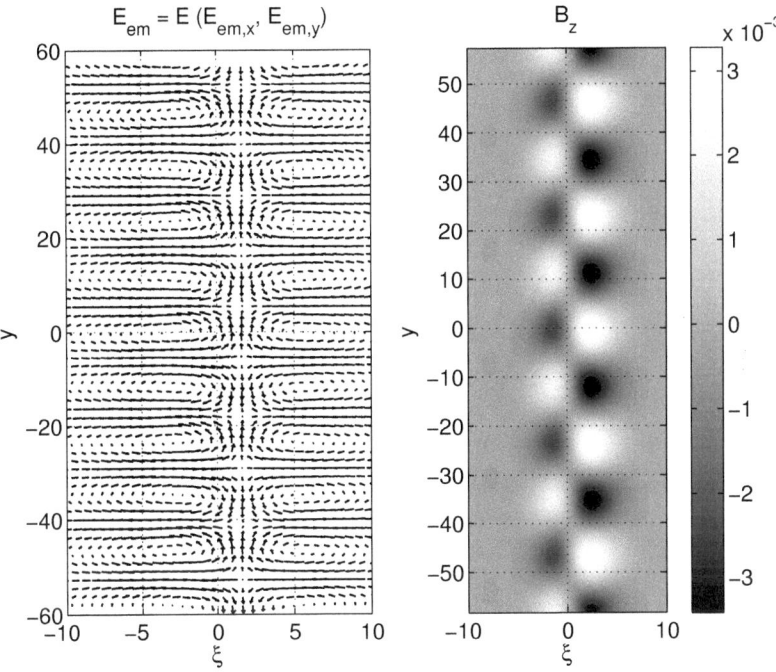

Figure 6.11: p-polarized field components of a perturbation for a soliton with $p = 0$ and $V = 0.1$. The wave-number of the perturbation is $k_\perp = 0.27$. Left: Field line plot of the E_x and E_y components of the electric field $\mathbf{E}_{em}$. Right: Contour plot of the B_z.

6. Two-dimensional dynamics of relativistic solitons

We compute the electromagnetic energy stored in the s-polarized and the p-polarized part by Eq. 6.15. Figure 6.12 displays the evolution of energy in the s- and the p-polarized part.

We find that the energy of the s-polarized part of the perturbation is about one order of magnitude larger than that of the p-polarized part. Hence we state that the fastest growing perturbation for the circular polarized Bulanov-Farina solitons is dominantly s-polarized.

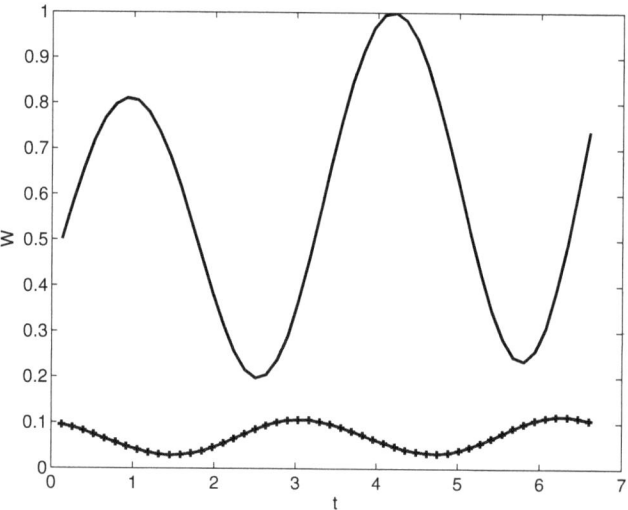

Figure 6.12: Energy W in the s-polarized (solid line) and the p-polarized (crossed line) part of the electromagnetic fields of the fastest growing perturbation for a $p = 0$ soliton with V=0.1 in arbitrary units. The evolution is shown for one period of the soliton.

6.4. Field structure

6.4.2 Nonlinear field structure

To determine the field structure of the nonlinear end state, we examine the fields that result from a nonlinear simulation of a perturbed soliton. We demonstrate the behavior for a $p = 0$ soliton with velocity $V = 0.1$. The soliton was perturbed by the linear mode that has been discussed in the previous section. The wave-number of the perturbation is $k_\perp = 0.27$. The initial amount of perturbation was 2%. Fig. 6.13 displays the s-polarized field components at t=104. At this point the transversal filamentation is completely developed, we see filaments separated by $\lambda_\perp = 2\pi/k_\perp$. The fields inside the filaments show periodic behavior and are quasi-stationary. The slow expansion of the density cavity due to the ponderomotive pressure of the trapped radiation would follow. Following the evolution for even longer time in the simulation is not yet feasible due to limitation of computational power.

When comparing the field structure to that of the linear mode we see that the geometry of the nonlinear fields are in very good accordance to those of the linear mode. The distance between the filaments matches the predictions from the linear regime.

What has changed in the nonlinear stage is the distribution of the electromagnetic energy. In the perturbation the s-polarized part carried the most energy of the electromagnetic fields. When we compute the electromagnetic energy in the nonlinear end state, we observe that this state is mostly p-polarized, see Fig. 6.14.

6.4.3 Relation to results from literature

The formation of solitons in relativistic laser-plasma interaction has been described in many places in literature [7, 8, 10, 36, 59, 71]. We want to relate the results for the nonlinear end state to those from experiments or PIC simulations.

Diagnostics for detection of solitons in experiments is able to probe the electric

6. Two-dimensional dynamics of relativistic solitons

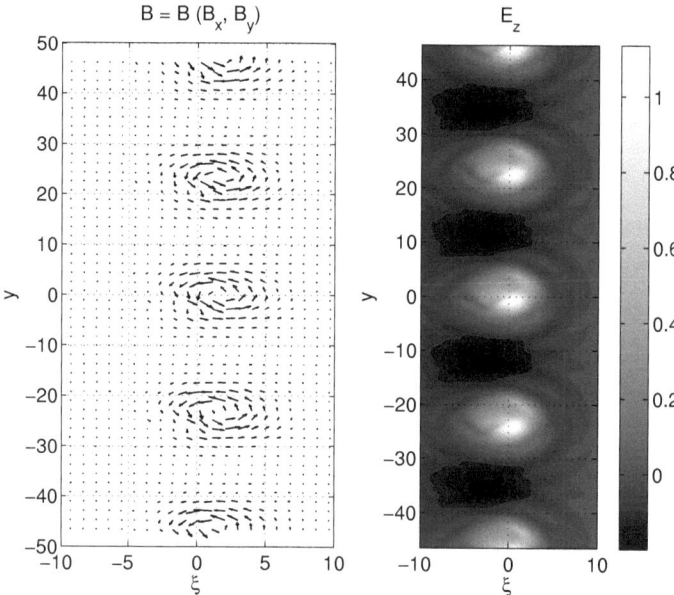

Figure 6.13: Magnetic and electric field of the s-polarized part of the nonlinear end state of a perturbed $p = 0$ soliton. For parameter details see text.

6.4. Field structure

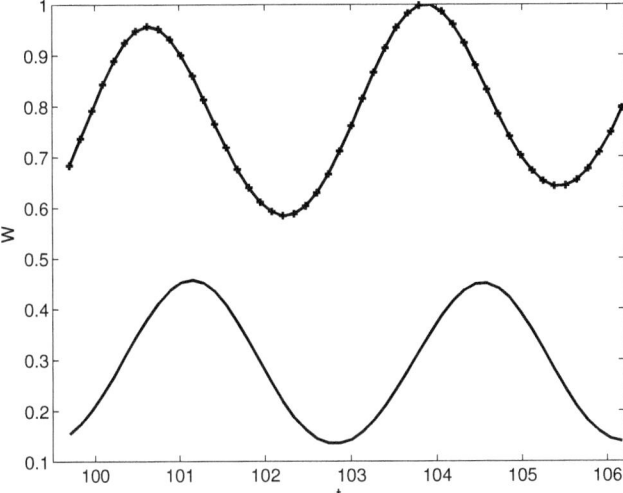

Figure 6.14: Energy W in the s- (solid line) and the p-polarized (crossed line) part of the electromagnetic fields of the nonlinear end state of perturbed $p = 0$ soliton with V=0.1 in arbitrary units. The initial perturbation was 2% of the fastest growing mode. The evolution is shown for one period of the soliton.

6. Two-dimensional dynamics of relativistic solitons

space charge fields which belong to a soliton [7, 8]. From these fields the spatial extension of the solitary structures can be determined. Radially symmetric structures are detected with sizes in the order of the collisionless electron skin depth $d_e = c/\omega_{pe}$. The radial size of the structures we observe in nonlinear 2D simulations matches these observations from the experiment. The polarization measurement of the soliton fields is a very recent development. It is found that the solitons created by a linear polarized laser have a dominant field component with the same polarization as the laser has. [36].

This finding is coherent with 2D PIC simulations, which show that solitons created by a s-polarized laser pulse have the same polarization as the driving laser [59, 71]. In these simulations, solitons are created by propagation of a laser pulse in an underdense plasma with varying density. At the top of a density profile the plasma is still underdense, but parts of the pulse become trapped due to frequency-downshift. Solitons are created and begin to expand very slowly.

In Ref. [10] the interaction of a circular polarized laser beam moving in an inhomogeneous plasma is discussed by a 2D PIC simulation. The pulse moves towards higher background density and creates a soliton. The field structure of the soliton is discussed and found to be of s-polarization. This prediction matches the results from our observations. From the transversal instability we obtain a solitary structure of the same dimensions and with a dominant polarization, too. The kind of polarization, p- or s-polarization, is different, but this may however depend on plasma parameters like for example the background density of the plasma. In our case the background density was homogeneous, whereas in Ref. [14] the density distribution was inhomogeneous.

Observation of the expanding behavior of the density cavity is beyond the computational possibilities of the Maxwell-fluid simulation at the current moment. Hence we are not able to confirm the rate of expansion of the cavity predicted by a snowplow model [69].

6.5 Instabilities in 3D

The study of soliton stability by nonlinear Maxwell-fluid simulations in 3D geometry is not yet feasible due to limitations of todays computers. Nevertheless the linear regime in 3D is accessible by the 2D formalism we developed in the previous sections.

The direction of propagation is x, so in general a perturbation in transversal direction has a wave-number $\mathbf{k}_\perp = (k_y, k_z)$ in the y, z-plane. In the previous sections where we discussed linear 2D stability we supposed $\bar{\mathbf{k}}_\perp = (\bar{k}_y, 0)$, since we were allowed to choose an arbitrary perpendicular direction.

To extend these results into 3D, we have to calculate transversal unstable modes for every angle ϕ, where ϕ is the angle between $\mathbf{k}_\perp$ and the y axis in the y, z-plane, i.e. $\phi = \arctan(k_z/k_y)$. Since the linear 2D simulation always demands $\bar{k}_z = 0$, we have to transform the results from this special system to systems that are rotated by ϕ to get perturbations that have a general wavenumber $\mathbf{k}_\perp = (k_y, k_z)$.

In order to see how the results from the system where $\mathbf{k}_\perp$ is purely in y direction connect to the system where $\mathbf{k}_\perp$ is an arbitrary vector in the y, z-plane, we have to rotate the 3D linearized equations (3.14)-(3.18) about the x direction by the angle ϕ. The rotation is such that $\mathbf{k}_\perp = (k_y, k_z) \to \bar{\mathbf{k}}_\perp = (\bar{k}_y, 0)$. All quantities have to be formulated in terms of the quantities in the rotated system, which are indicated by bars. Thus the rotation is given by

$$y = \bar{y}\cos\phi - \bar{z}\sin\phi, \qquad (6.16)$$

$$z = \bar{y}\sin\phi + \bar{z}\cos\phi. \qquad (6.17)$$

The perpendicular components of the vector quantities are transformed by the same

6. Two-dimensional dynamics of relativistic solitons

scheme, for example the components of vector potential **A**

$$A_y = \bar{A}_y \cos\phi - \bar{A}_z \sin\phi\,, \tag{6.18}$$

$$A_z = \bar{A}_y \sin\phi + \bar{A}_z \cos\phi\,. \tag{6.19}$$

Transformation of the system (3.14)-(3.18) by this rotation and rewriting all equations for the rotated quantities leads to a description that is equivalent to the 2D linearized system (6.6)-(6.9), but where the unperturbed state is now rotated by the angle ϕ. Thus we have to add a constant phase $e^{i\phi}$ to the unperturbed state to grow the most unstable mode with the existing 2D linear code for a system that is rotated by ϕ in the y, z plane. All simulations show that the growth rate of the most unstable mode does not depend on the initial phase of the unperturbed state, hence the growth rate does not depend on the direction of $\mathbf{k}_\perp$, but only on the magnitude of $\mathbf{k}_\perp$. As the whole system behaves linear, it is now possible to superpose modes with different $\mathbf{k}_\perp$ to construct a 3D mode that grows exponentially within the linear regime.

However, in full 3D geometry the nonlinear evolution of these modes may be completely different from the 2D case. It is therefore not possible, to strictly transfer the observations from 2D nonlinear simulations into 3D geometry.

7 Conclusion

In the present work the stability of solitary waves including wave-breaking criteria has been studied. The stability of relativistic solitons is of importance as it gives insights in the creation and evolution of relativistic plasma structures that can trap laser radiation. Wave-breaking criteria determine the stability of plasma waves, which is of fundamental importance for plasma-based particle accelerators.

The interaction of plasma and a laser pulse was described by Maxwell-fluid equations. A general numerical method for linear stability analysis was developed and applied to different soliton solutions in 1D and 2D geometry.

At first, different types of solitons were examined in a 1D model. In 1D geometry all quantities depend spatially on just one coordinate (eventhough they may be three-dimensional by themself). Within the approximation of a static ion background and cold electrons standing pre-soliton solutions are known. By a numerical linear stability analysis, we found them to be stable. Their transition from pre- to post-solitons was demonstrated by allowing the initially homogeneous ion density to vary and react to the presence of the soliton. The creation and further development of a density cavity was shown for this case.

Linear polarized pre-solitons were derived in the limit of small laser amplitudes, small plasma density variations and the assumption of a fixed ion background. We showed that these solutions are unstable. The analytical stability criterion for these solitons that was previously used in literature was demonstrated not to apply here.

7. Conclusion

To investigate stability of solitons on time-scales larger than $t_i = \sqrt{m_i/m_e} \approx 40$, the ions have to be included in the initial soliton solution. We performed a stability analysis for solitons in cold and in warm plasma, both for circular polarized lasers [53]. The standing soliton solutions for a warm electron-ion plasma only show perturbations that grow according to a power-law in time and not exponentially. These modes were identified as phasor-modes and not as an instability, so the solitons can be considered stable.

In cold plasma no standing soliton solutions can exist in the case of mobile ions. The solitons always have to have a certain minimal velocity. In this regime we studied stationary Farina-Bulanov solitons. The envelope of the vector potential can have different numbers $p = 0, 1, 2, ...$ of nodes. We found that solitons with node number $p = 0$ show no physically relevant longitudinal instabilities. All solitons with node numbers $p \geq 1$ were demonstrated to suffer from instability on the ion time-scale. Two kinds of longitudinal instabilities were found for $p \geq 1$ solitons, a purely growing one and an oscillatory growing one. Below a certain velocity the instability is purely growing in time, above this velocity additional oscillations in time appear.

The nonlinear evolution of a perturbation for unstable Farina-Bulanov solitons showed the excitation of plasma waves in the wake of the soliton. Eventually we observed breaking of these electrostatic waves. In order to gain insight into the stability of wake-fields we studied general criteria for wake-field excitation and wave-breaking.

Criteria for wave-breaking in different scenarios are important, because on the one hand they give the maximum electrical field that a plasma can sustain and on the other hand they make predictions about the life-time of the wake-field. Both is important for particle acceleration schemes using wake-fields. Electrostatic plasma waves with frequency ω_{pe} are known to break within the first oscillation if their amplitude is sufficiently large. By a relativistic Lagrangian coordinate description of the electron fluid, the possibility of breaking in the presence of an inhomogeneous background density has already been shown. We succeeded in showing up another mechanism for wave breaking in the relativistic regime [54]. In Lagrangian description the electrical field amplitude E_L at the place of the fluid element is governed by a nonlinear oscillator equation. The nonlinearity is introduced by the relativistic γ factor. We showed that this nonlinearity, however small it may be, will eventually lead to wave-breaking. The time of breaking T is in general larger than $1/\omega_{pe}$, and we derived an expression for an estimate of T. Hence even in the case of homogeneous background density an electrostatic wave will always break due to relativistic effects.

The Lagrangian model is only suited for electrostatic waves, it cannot give a description of wave breaking at positions where the laser is present. Furthermore it does not strictly hold for times that are in the order of the ion time t_i, but by supporting additional Maxwell-fluid simulations we showed that mobile ions only reduce the time it takes the wave to break, hence the Lagrangian model gives an appropriate upper limit.

So far we treated the stability of solitons in 1D geometry. To reduce the general 3D Maxwell-fluid model to a 1D model, we assumed that all quantities are transversely constant. To investigate the influence of transversal perturbations, we had to allow for variation of all quantities in a transversal direction. This reduces the 3D model to a 2D model, which may even be physically three-dimensional if an appropriate symmetry is assumed. With the 2D model we studied the stability of Bulanov-Farina solitons which are transversely (to the propagation direction) perturbed. For stability analysis we employed the same method as for 1D stability, which includes determining the structure of the fastest growing perturbation.

We found that the longitudinally stable $p = 0$ solitons are transversely unstable. The growth rate depends on the wave-number $k_\perp$ of the transversal perturbation and is non-zero in an interval $0 \leq k_\perp \leq k_{\perp c}$. Perturbations with a wave-number above

7. Conclusion

a certain cut-off $k_{\perp c}$ do not grow exponentially within the linearized model. The maximum growth rate Γ for the transversal instability is about $\Gamma \approx 0.1$, which allows the growth of such modes in a physically relevant time. The maximum growth rate depends on the soliton velocity V. The faster the soliton, the lower the maximum growth rate for transversal instability.

For the $p \geq 1$ solitons we found similar effects of transversal instability as for $p = 0$ solitons. The qualitative dependence of the growth rate from the wave-number of the perturbation is the same as for $p = 0$ solitons. In general the maximum growth rate increases with increasing node number (at fixed soliton velocity V). For fixed p number, the maximum growth rate increases as the soliton velocity decreases. Since solitons with $p \geq 1$ are longitudinally unstable, a comparison between longitudinal and transversal instability was carried out. The transversal instability is always dominant over the longitudinal instability. For $p = 1$ solitons the ratio of the growth rates (transversal growth rate above longitudinal) is at least a factor of two, the maximum ratio is 14 in the linear regime.

By nonlinear simulations we demonstrated, that the nonlinear evolution of the perturbations resulted always in strongly localized laser filaments which produce a plasma density depression. The transversal wave-number of the nonlinear laser filament structure is still that of the linear mode. As we observed no wave-vector cascade (at least in the wave-vector domain that the simulations allow us to follow) we conclude that Haken's slaving principle determines the transition into the nonlinear regime, which says that the linear most unstable mode will slave all other modes and will show up in the topology of the nonlinear end state.

To demonstrate the transversal instability for solitons which are not perturbed by the exact initial distribution of the most unstable mode, we showed that it arises from noise. Nonlinear simulations were carried out with solitons perturbed by random initial noise. A clear transversal instability arises with a wave number matching exactly the most unstable linear mode.

To get a more physical understanding of the 2D perturbation and the nonlinear end state, we studied their field structure. The perturbation consists of two coupled parts of laser radiation, a s-polarized and a p-polarized part. The s-polarized part contained the largest part of the energy of the perturbation, about an order of magnitude more than the p-polarized part. Hence we state that the fastest growing perturbation of a circular polarized Bulanov-Farina soliton is always dominantly s-polarized. In the nonlinear end state however, we see that the laser filaments that are created from the initial perturbation are dominantly p-polarized. Due to nonlinear effects the system distributes more energy to the p-polarized fields than to the s-polarized.

Finally we discussed how the 2D case can be generalized to construct linear 3D modes with known growth rate. The study of the nonlinear evolution of these modes in 3D has to be treated by upcoming projects.

7. Conclusion

A Appendix

A.1 Stability of invariant sets

We introduce a metric $d(y_1, y_2)$ to measure the distance between two states or functions y_1, y_2. If the metric is induced by a norm

$$d(y_1, y_2) = \sqrt{||y_1 - y_2||}, \tag{A.1}$$

the distance between unperturbed and perturbed state can be interpreted as the size of the perturbance.

Let p be a perturbed state that originates from an unperturbed state u with a small perturbation added. A time-independent unperturbed state u is stable, if there is a $\delta > 0$ for every $\epsilon > 0$, such that for all times $t \geq t_0$, for arbitrary t_0 from

$$||p(t_0) - u|| < \delta, \tag{A.2}$$

follows

$$||p(t) - u|| < \epsilon. \tag{A.3}$$

This means that the distance between unperturbed and perturbed state stays small for all times if the initial distance was sufficiently small.

If the state u is time-dependent one possible way to define stability is to allow an

A. Appendix

explicit time dependence of u in the above definition and measure the distance instantaneously. This definition would be too strict for the stability analysis of solitons. If we consider two solitons with slightly different amplitudes they will separate arbitrarily far from each other since their velocity depends on their amplitude. Nevertheless this is not an instability as an instantaneous measure of the distance between the solitons would suggest.

A more general approach is the stability of invariant sets, which takes account of e.g. translations or phase differences. Therefore we consider the whole set S of unperturbed trajectories T as an unperturbed state and measure the distance between this set of trajectories and the perturbed state p

$$d(p, S) = \inf_{u \in S} ||p - u||. \tag{A.4}$$

If for every $\epsilon > 0$ there exists a $\delta > 0$ such that for all times t_0 with $t \geq t_0$

$$d(p(t_0), S) < \delta \Rightarrow d(p(t), S) < \epsilon. \tag{A.5}$$

we call the set S stable.

B Appendix

B.1 Analytical stability criteria

The N-Theorem for the nonlinear Schrödinger equation

For localized solitary wave solutions of the nonlinear Schrödinger equation

$$i\partial_t \psi + \partial_x^2 \psi + U'\psi - \beta(\partial_x^2 |\psi|^2)\psi = 0 \tag{B.1}$$

where ψ is a complex envelope, U is a real valued potential and β is a positive constant, the region of stability can be determined by the N-Theorem or Vakitov-Kolokolov criterion.

Let us suppose that a localized and quasi-stationary solution exists. We write the latter in the form

$$\psi_s = G(x, \eta_s^2) \exp(i\eta_s^2 t), \tag{B.2}$$

then G is given by

$$-\eta_s^2 G + \partial_x^2 G + U'(G^2)G - \beta(\partial_x^2 G^2)G = 0, \tag{B.3}$$

the exact form of G is not important at this point.

B. Appendix

It is possible to integrate (B.3) once, after multiplication by $\partial_x G$, we get

$$(\partial_x G)^2 (1 - 2\beta G^2) - \eta_s^2 G^2 + U(G^2) = 0. \tag{B.4}$$

By integration of (B.4) G follows for a given U by integration.

The equation (B.1) has multiple constants of motion, we will discuss two here. Multiplication of (B.1) with ψ^*, subtracting the complex conjugate and integrate over space, we find

$$\partial_t N = 0, \tag{B.5}$$

where the quasiparticle number

$$N = \int_{-\infty}^{\infty} |\psi|^2 \mathrm{d}x \tag{B.6}$$

has been introduced.

Multiplication of both sides of Eq. (B.1) with $\partial_t \psi^*$, adding the complex conjugate, and integrating over space, we get the energy conservation

$$\partial_t E = 0 \tag{B.7}$$

where

$$E = \int_{-\infty}^{\infty} \left(|\partial_x \psi|^2 - U(|\psi|^2 - \frac{\beta}{2}(\partial_x |\psi|^2)^2 \right). \tag{B.8}$$

We now perturb the soliton G in the form

$$\psi = (G + a + ib) \exp(i\eta^2 t) \tag{B.9}$$

and study the evolution of the perturbations a and b. We will restrict ourself to the

B.1. Analytical stability criteria

linear evolution of the perturbation. By inserting (B.9) into (B.1), we obtain

$$\partial_t a = H_+ b \tag{B.10}$$

and

$$\partial_t b = -H_- a, \tag{B.11}$$

where the Schrödinger operators H_+ and H_- are defined as

$$H_+ = -\partial_x^2 + \eta_s^2 - U' + \beta(\partial_x^2 G^2) \tag{B.12}$$

$$H_- = H_+ - 2G^2 U'' + 2\beta G \partial_x^2 G. \tag{B.13}$$

From (B.12) and (B.10) follows that $\partial_t \langle a|G\rangle = 0$, hence it is sufficient to treat perturbations with $\langle a|G\rangle = 0$. We assume G to be a bell-shaped soliton (so G has no nodes). Then H_+ is positive semidefinite and H_- has one negative eigenvalue, since $\partial_x G$ has one node.

To determine the maximum growth rate Γ of an instability a variational principle can be derived. We consider functions ξ which are perpendicular to the kernel of H_+. In the following the components of ξ parallel to G vanish, i.e. $\langle \xi|G\rangle = 0$. Using the results from Ref. [47] and [48] we get

$$\Gamma^2 = \sup_{\substack{\xi \\ \langle \zeta|G\rangle=0}} \frac{-\langle \xi|H_-|\xi\rangle}{\langle \xi|H_+^{-1}|\xi\rangle}. \tag{B.14}$$

To derive a criterion for instability we construct a ξ (with $\langle \xi|G\rangle = 0$) such that $\langle \xi|H_-|\xi\rangle < 0$. Let us consider

$$\xi = \langle \xi_-|G\rangle H_-^{-1} G - \langle G|H_-^{-1}|G\rangle \xi_- \tag{B.15}$$

B. Appendix

where $H_-^{-1}G$ is defined by

$$H_-^{-1}G = -\frac{\partial G}{\partial \eta_s^2} \tag{B.16}$$

and ξ_- is an arbitrary function with $\langle \xi_- | H_- | \xi_- \rangle < 0$. Such a ξ_- always exists since H_- has a negative eigenvalue.

Using (B.15) we find

$$\langle \xi | H_- | \xi \rangle = -\langle G | H_-^{-1} | G \rangle \left(\langle \xi_- | G \rangle^2 - \langle G | H_-^{-1} | G \rangle \langle \xi_- | H_- | \xi_- \rangle \right). \tag{B.17}$$

Thus instability can occur provided

$$\langle G | H_-^{-1} | G \rangle > 0, \tag{B.18}$$

and from Eq. (B.16) we see that this implies

$$\frac{\partial N_s}{\partial \eta_s^2} < 0 \tag{B.19}$$

where N_s is the soliton quasiparticle number

$$N_s = \int_{-\infty}^{\infty} G^2 \, \mathrm{d}x. \tag{B.20}$$

This is a suitable criterion for instability. We cite that in the opposite case the solitons are stable [88, 50].

The Q-Theorem for the Klein-Gordon equation

To present and prove the Q-Theorem for stability of solutions of Klein-Gordon equations we will make use of the notation used in Ref. [45]. We start with the Lagrangian

B.1. Analytical stability criteria

density
$$\mathcal{L} = \partial_\mu \phi \partial^\mu \phi^* - U(|\phi|^2), \tag{B.21}$$

where ϕ is in general a complex (charge) field. The potential U is real valued and is field dependent.

The equation of motion follow from the principle of least action

$$\delta \int \mathrm{d}^4 \mathcal{L} = 0. \tag{B.22}$$

The variation leads to the Euler-Lagrange equation

$$\partial_\mu \phi \partial^\mu \phi + \frac{\partial U}{\partial |\phi|^2} \phi = 0. \tag{B.23}$$

We will suppose the metric to be defined in such a way that $\partial_\mu \partial^\mu = \partial_t^2 - \nabla$.

The system conserves energy E and charge Q which are given by

$$E = \int_{-\infty}^{\infty} \left(|\partial_t \phi|^2 + |\nabla \phi|^2 + U(|\phi|^2) \right) \mathrm{d}^3 x, \tag{B.24}$$

$$Q = -\mathrm{i} \int_{-\infty}^{\infty} \left(\phi \frac{\partial \phi^*}{\partial t} - \phi^* \frac{\partial \phi}{\partial t} \right) \mathrm{d}^3 x. \tag{B.25}$$

We suppose the solitons to be quasi-stationary and of the form

$$\phi_s = \phi_s \exp(-\mathrm{i}\mu_s t), \tag{B.26}$$

where $\phi_s \to 0$ for $x \to \pm\infty$.

The solitons have the charge

$$Q_s = 2\mu_s \int_{-\infty}^{\infty} \phi_s^2 \mathrm{d}^3 x \tag{B.27}$$

B. Appendix

and obey the equation

$$-\partial_x^2 \phi_s - \mu_s \phi_s + \frac{\partial U}{\partial \phi_s^2} \phi_s = 0. \tag{B.28}$$

We will prove that the solitons are stable if

$$\frac{\mathrm{d}Q_s}{\mathrm{d}\mu_s} < 0. \tag{B.29}$$

The stability criterion will be proven in 1D geometry, i.e. all quantities depend only on one spatial coordinate, say x. In order to discuss stability properties, we will construct a Ljapunov function out of the energy functional E and the energy E_s of the zeroth order soliton solution in the following manner

$$L = E - E_s. \tag{B.30}$$

Since E is a constant of motion, $\mathrm{d}L/\mathrm{d}t = 0$ is always satisfied. It remains to show that, in the neighborhood of the stationary point, L is positive. When L is positive it can be estimated in terms of the norm. The norm which will be used is the Sobolev norm

$$||F||^2 = \int \left((\partial_x F)^2 + F^2 \right) \mathrm{d}x. \tag{B.31}$$

We will discuss the existence of a lower bound of L in terms of this norm.

By using Schwarz inequality

$$\int (\phi_t \phi_t^*) \, \mathrm{d}x \int (\phi \phi^*) \, \mathrm{d}x \geq |\int (\phi \phi_t^*) \, \mathrm{d}x|^2 \geq \left\{ \mathrm{Im} \int (\phi \phi_t^*) \, \mathrm{d}x \right\}^2 \tag{B.32}$$

it is possible to show that

$$E \geq A \tag{B.33}$$

B.1. Analytical stability criteria

where

$$A = \frac{1}{2} \int \left(|\partial_x \phi|^2 + U(|\phi|^2) \right) dx + \frac{\mu^2}{4} S \tag{B.34}$$

$$S = 2 \int \phi^* \phi \, dx \,, \tag{B.35}$$

and $\mu = Q/S$. Note $A_s = E_s$, hence we investigate the quantity

$$R = A - A_s \,, \tag{B.36}$$

which is smaller than L.

We take the first variation of R and assume $\delta Q = 0$ and get

$$\delta R = \frac{1}{2} \int \left(\left(-\partial_x^2 \phi_s + \frac{\partial U}{\partial \phi_s^2} \phi_s^2 - \mu_s \phi_s \right) \delta \phi^* \right) dx + \text{c.c.} \tag{B.37}$$

Due to (B.28), $\delta R = 0$.

To show that R has a local minimum for the soliton state, we consider the second variation. We get

$$\delta^2 R = \int \left(\partial_x \delta\phi \partial_x \delta\phi^* + \frac{1}{2} \frac{\partial^2 U}{\partial (\phi_s^2)^2} [\phi_s \delta\phi^* + \delta\phi \, \phi_s^*]^2 + \frac{\partial U}{\partial \phi_s^2} \delta\phi \delta\phi^* \right) dx$$
$$- \mu_s^2 \int (\delta\phi \delta\phi^*) \, dx + \frac{\mu_s^3}{2 Q_s} (\delta S)^2 \tag{B.38}$$

We denote the perturbed states by

$$\phi = (\phi_s + a + ib) \exp(-i\mu_s t) \,. \tag{B.39}$$

B. Appendix

Inserting into (B.38) we get

$$\delta^2 R = \int (aH_-a + bH_+b)\,\mathrm{d}x + 4\mu_s^2 \frac{(\int(\phi_s a)\mathrm{d}x)^2}{\int \phi_s^2 \mathrm{d}x} \tag{B.40}$$

where

$$H_+ = -\partial_x^2 + \frac{\partial U}{\partial \phi_s^2} - \mu_s^2, \tag{B.41}$$

$$H_- = -\partial_x^2 + \frac{\partial U}{\partial \phi_s^2} - \mu_s^2 + 2\phi_s^2 \frac{\partial^2 U}{\partial (\phi_s^2)^2}. \tag{B.42}$$

Using bra and ket notation, we rewrite Eq. (B.40) as

$$\delta^2 R = \langle a|H_-|a\rangle + \langle b|H_+|b\rangle + 4\mu_s^2 \frac{\langle a|\phi_s\rangle^2}{\langle \phi_s|\phi_s\rangle}. \tag{B.43}$$

The operator H_+ is positive semidefinite for single humped solutions ϕ_s, since $H_+ \phi_s = 0$. To discuss the definiteness properties of $\delta^2 R$, we have to investigate the terms $\langle a|H_-|a\rangle$ and $\langle a|\phi_s\rangle^2/\langle \phi_s|\phi_s\rangle$.

We make use of Ref. [47] and calculate

$$\sup_{\psi} \frac{-\langle \psi|H_-|\psi\rangle}{\langle \psi|\phi_s\rangle^2} = \frac{1}{-\langle \phi_s|H_-^{-1}|\phi_s\rangle} \tag{B.44}$$

for operators H_- with one and only one negative eigenvalue, we obtain

$$\langle a|H_-|a\rangle \geq \frac{\langle a|\phi_s\rangle^2}{\langle \phi_s|H_-^{-1}|\phi_s\rangle}. \tag{B.45}$$

Furthermore we have

$$H_-^{-1}\phi_s = \frac{1}{2\mu_s}\frac{\partial}{\partial \mu_s}\phi_s \tag{B.46}$$

B.1. Analytical stability criteria

and thereby
$$\langle \phi_s | H_-^{-1} | \phi_s \rangle = -\frac{Q_s}{8\mu_s} + \frac{1}{8\mu_s}\frac{\mathrm{d}Q_s}{\mathrm{d}\mu_s}. \tag{B.47}$$

Combining (B.43)-(B.47), we obtain

$$\delta^2 R \geq \left(\frac{8\mu_s^3}{Q_s} - \frac{1}{(Q_s/8\mu_s^3) - (1/8\mu_s^3)(\mathrm{d}Q_s/\mathrm{d}\mu_s)}\right) \langle a | \phi_s \rangle^2. \tag{B.48}$$

This is larger than zero for
$$\frac{\mu_s}{Q_s}\frac{\mathrm{d}Q_s}{\mathrm{d}\mu_s} < 0, \tag{B.49}$$

which is the stability criterion we wanted to prove.

Let us make some notes on the restriction that the perturbed states have the same charge as the unperturbed ground state which is considered, i.e. $\delta Q = 0$. Introducing perturbations which change the charge, so $\Delta Q = Q - Q_s \lesseqgtr 0$, we might come up with a larger instability region. The definiteness of L could be recalculated without this assumption, or we may use a more direct argument.

Physically, we are investigating the stability of form. Suppose we are examining a state with conserved charge Q that is different from the charge of the unperturbed soliton Q_s. Then it is possible to find another soliton ($\bar{s}$), having the same charge as the perturbed state $Q_{\bar{s}}$. The perturbation under consideration is now stable with respect to this new soliton of equal charge. Since we are investigating the stability of form, the distance between the two solitons is constant. By employing the triangle inequality, one can show that the perturbed state (with charge Q) is also stable with respect to the unperturbed state s (with charge Q_s).

B. Appendix

Bibliography

[1] A. I. Akhiezer and R. V. Polovin, *Theory of wave motion of an electron plasma*, Sov. Phys. JETP **3** (1956) 696.

[2] F. Amiranoff, A. Antonietti, P. Audebert, D. Bernard, B. Cros, F. Dorchies, J. C. Gauthier, J. P. Geindre, G. Grillon, F. Jacquet, G. Matthieussent, J. R. Marquès, P. Mine, P. Mora, A. Modena, J. Morillo, F. Moulin, Z. Najmudin, A. E. Specka, and C. Stenz, *Laser particle acceleration: beat-wave and wakefield experiments*, Plasma Phys. Control. Fusion **38** (1996) 295.

[3] F. Amiranoff, D. Bernard, B. Cros, F. Dorchies, F. Jacquet, V. Malka, J. R. Marques, G. Matthieussent, P. Mine, A. Modena, J. Morillo, and Z. Najmudin, *The laser wakefield acceleration experiment at Ecole Polytechnique*, Nuclear Instruments and Methods in Physics Research Section A **410** (1998) 364–366.

[4] S. Banerjee, S. Sepke, R. Shah, A. Valenzuela, A. Maksimchuk, and D. Umstadter, *Optical deflection and temporal characterization of an ultrafast laser-produced electron beam*, Physical Review Letters **95**(3) (2005).

[5] V. Berezhiani and I. Murusidze, *Interaction of highly relativistic short laser pulses with plasmas and nonlinear wake-field generation*, Physica Scripta **45**(2) (1992) 87–90.

Bibliography

[6] V. I. Berezhiani, S. M. Mahajan, Z. Yoshida, and M. Ohhashi, *Self-trapping of strong electromagnetic beams in relativistic plasmas*, Phys. Rev. E **65**(4) (2002) 047402.

[7] M. Borghesi, S. Bulanov, D. Campbell, R. J. Clarke, T. Esirkepov, M. Galimberti, L. A. Gizzi, A. J. MacKinnon, N. Naumova, F. Pegoraro, H. Ruhl, A. Schiavi, and O. Willi, *Macroscopic Evidence of Soliton Formation in Multiterawatt Laser-Plasma Interaction*, Phys. Rev. Lett. **88** (2002) 135002.

[8] M. Borghesi, D. Campbell, A. Schiavi, M. G. Haines, O. Willi, A. J. MacKinnon, P. Patel, L. A. Gizzi, M. Galimberti, E. L. Clark, F. Pegoraro, H. Ruhl, and S. Bulanov, *Electric field detection in laser-plasma interaction experiments via he proton imaging technique*, Plasma Phys. **9** (2002) 214–2220.

[9] T. J. Bridges, *On the susceptibility of bright nonlinear Schrödinger solitons to long-wave transverse instability*, Proceedings of the Royal Society A: Mathematical, Physical and Engineering Sciences **460** (2004) 2605–2615.

[10] S. Bulanov, F. Califano, T. Esirkepov, K. Mima, N. Naumova, K. Nishihara, F. Pegoraro, Y. Sentoku, and V. A. Vshivkov, *Generation of subcycle relativistic solitons by super intense laser pulses in plasma*, Physica D **152-153** (2001) 682–693.

[11] S. Bulanov, T. Esirkepov, N. Naumova, F. Pegoraro, and V. A. Vshivkov, *Solitonlike Electromagnetic Waves behind a Superintense Laser Pulse in a Plasma*, Phys. Rev. Lett. **82** (1999) 3440–3443.

[12] S. Bulanov and F. Pegoraro, *Stability of a mass accreting shell expanding in a plasma*, Phys. Rev. E **65** (2002) 066405.

Bibliography

[13] S. V. Bulanov, *New epoch in the charged particle acceleration by relativistically intense laser radiation*, Plasma Phys. Control. Fusion **48** (2006) 29.

[14] S. V. Bulanov, F. Califano, G. I. Dudnikova, T. Z. Esirkepov, I. N. Inovenkov, F. F. Kamenets, T. V. Liseikina, M. Lontano, K. Mima, N. M. Naumova, K. Nishihara, F. Pegoraro, H. Ruhl, A. S. Sakharov, Y. Sentoku, V. A. Vshivkov, and V. V. Zhakhovskii, *Reviews of Plasma Physics*, volume 22, chapter Relativistic Interaction of Laser Pulses with Plasmas, pages 227–335, Kluwer Academic/Plenum Publishers, 2001.

[15] S. V. Bulanov, F. Pegoraro, and A. M. Pukhov, *Two-Dimensional Regimes of Self-Focusing, Wake Field Generation, and Induced Focusing of a Short Intense Laser Pulse in an Underdense Plasma*, Phys. Rev. Lett. **74**(5) (1995) 710–713.

[16] S. V. Bulanov, F. Pegoraro, A. M. Pukhov, and A. S. Sakharov, *Transverse-wake wave breaking*, Phys. Rev. Lett. **78**(22) (1997) 4205–4208.

[17] S.-Y. Chen, M. Krishnan, A. Maksimchuk, R. Wagner, and D. Umstadter, *Detailed dynamics of electron beams self-trapped and accelerated in a self-modulated laser wakefield*, Physics of Plasmas **6**(12) (1999) 4739–4749.

[18] T. P. Coffey, *Breaking of Large Amplitude Plasma Oscillations*, Physics of Fluids **14**(7) (1971) 1402–1406.

[19] J. M. Dawson, *Nonlinear electron oscillations in a cold plasma*, Phys. Rev. **113** (1959) 383.

[20] C. D. Decker and W. B. Mori, *Group velocity of large amplitude electromagnetic waves in a plasma*, Phys. Rev. Lett. **72**(4) (1994) 490–493.

[21] R. J. England, J. B. Rosenzweig, and N. Barov, *Plasma electron fluid motion and wave breaking near a density transition*, Phys. Rev. E **66** (2002) 016501.

Bibliography

[22] E. Esarey and M. Pilloff, *Trapping and acceleration in nonlinear plasma waves*, Physics of Plasmas **2**(5) (1995) 1432–1436.

[23] T. Esirkepov, F. F. Kamenets, and N. Bulanov, S.; Naumova, *Low-frequency relativistic electromagnetic solitons in collisionless plasma*, JETP Lett. **68** (1998) 36–41.

[24] T. Esirkepov, K. Nishihara, and F. Bulanov, S.; Pegoraro, *Three-Dimensional Relativistic Electromagnetic Subcycle Solitons*, Phys. Rev. Lett. **89** (2002) 275002.

[25] D. Farina and S. Bulanov, *Dynamics of Relativistic Solitons*, Plasma Phys, Control. Fusion **47** (2005) A73–A80.

[26] D. Farina and S. V. Bulanov, *Relativistic Electromagnetic Solitons in the Electron-Ion Plasma*, Phys. Rev. Lett. **86**(23) (2001) 5289–5292.

[27] J. Faure, Y. Glinec, A. Pukhov, S. Kiselev, S. Gordienko, E. Lefebvre, J.-P. Rousseau, F. Burgy, and V. Malka, *A laser-plasma accelerator producing monoenergetic electron beams*, Nature **431** (2004) 541.

[28] C. G. R. Geddes, C. Toth, J. van Tillborg, E. Esarey, C. B. Schroeder, D. Bruhwller, C. Nieter, J. Cary, and W. P. Leemans, *High-quality electron beams from a laser wakefield accelerator using plasma-channel guiding*, Nature **431** (2004) 538.

[29] E. Gerstner, *Extreme Light*, Nature **446** (2007) 16–18.

[30] J. Goodman, T. Hou, and E. Tadmor, *On the stability of the unsmoothed Fourier method for hyperbolic equations*, Numer. Math. **67** (1994) 93 – 129.

[31] L. M. Gorbunov and V. I. Kirsanov, *Excitation of plasma waves by an electromagnetic wave packet*, Zh. Eksp. Teor. Fiz. **93** (1987) 509.

Bibliography

[32] L. Hadžievski, M. S. Jovanović, M. M. Škorić, and K. Mima, *Stability of one-dimensional electromagnetic solitons in relativistic laser plasmas*, Phys. Plasmas **9** (2002) 2569.

[33] L. Hadžievski, A. Mančić, and M. Škorić, *Moving Weakly Relativistic Electromagnetic Solitons in Laser-Plasmas*, ArXiv Physics e-prints **physics/0410209** (2004).

[34] H. Haken, *Slaving principle revisited*, Physica D **97**(1-3) (1996) 95–103.

[35] D. Kaganovich, A. Ting, D. F. Gordon, R. F. Hubbard, T. G. Jones, A. Zigler, and P. Sprangle, *First demonstration of a staged all-optical laser wakefield acceleration*, Phys. Plasmas **12** (2005) 100702.

[36] M. Kando, *in preparation*, (2008).

[37] C. Karle, J. Schweitzer, M. Hochbruck, E. W. Laedke, and K. H. Spatschek, *Numerical solution of nonlinear wave equations in stratified dispersive media*, J. Comp. Phys. **216** (2006) 138–152.

[38] T. Katsouleas and W. B. Mori, *Wave-Breaking Amplitude of Relativistic Oscillations in a Thermal Plasma*, Phys. Rev. Lett. **61**(1) (1988) 90–93.

[39] P. K. Kaw, A. Sen, and T. Katsouleas, *Nonlinear 1D Laser Pulse Solitons in a Plasma*, Phys. Rev. Lett. **68**(21) (1992) 3172–3175.

[40] Y. S. Kivshar and D. E. Pelinovsky, *Self-focusing and transverse instabilities of solitary waves*, Physics Reports **331** (2000) 117–195.

[41] R. Kodama, Y. Sentoko, Z. L. Chen, G. R. Kumar, S. P. Hatchett, Y. Toyama, T. E. Cowan, R. R. Freeman, J. Fuchs, Y. Izawa, M. H. Key, Y. Kitagawa,

Bibliography

K. Kondo, T. Matsuoka, H. Nakamura, M. Nakatsutsumi, P. A. Norreys, T. Norimatsu, R. A. Snavely, R. B. Stephens, M. Tampo, K. A. Tanaka, and T. Yabuuchi, *Plasma devices to guide and collimate a high density of MeV electrons*, Nature **432** (2004) 1005.

[42] V. A. Kozlov, A. G. Litvak, and E. V. Suvorov, Sov. Phys. JETP **49** (1979) 75.

[43] J. Krall, A. Ting, E. Esarey, and P. Sprangle, *Enhanced acceleration in a self-modulated-laser wake-field accelerator*, Phys. Rev. E **48**(3) (1993) 2157–2161.

[44] K. Krushelnik, E. L. Clark, F. N. Beg, A. E. Dangor, Z. Najmudin, P. A. Norreys, M. Wei, and M. Zepf, *High intensity laser-plasa sources of ions-physics and future applications*, Plasma Phys. Control. Fusion **47** (2005) 451–463.

[45] E. Laedke and K. Spatschek, *Advances in Nonlinear Waves*, chapter The Q-Theorem for Charged Solitons, Pittman, 1984.

[46] E. W. Laedke, *Zur nichtlinearen Stabilität solitärer Wellen in Plasmen*, Dissertation, Universität Essen, 1978.

[47] E. W. Laedke and K. H. Spatschek, *Complementary energy principles in dissipative fluids*, Journal of Mathematical Physics **23**(3) (1982) 460–463.

[48] E. W. Laedke and K. H. Spatschek, *Growth rates of bending KdV solitons*, Journal of Plasma Physics **28** (1982) 469–484.

[49] E. W. Laedke and K. H. Spatschek, *Thresholds and growth times of some nonlinear field equations*, Physica **5D** (1982) 227–242.

[50] E. W. Laedke, K. H. Spatschek, and L. Stenflo, *Evolution theorem for a class of perturbed envelope soliton solutions*, Journal of Mathematical Physics **24**(12) (1983) 2764–2769.

Bibliography

[51] W. Leemans, B. Nagler, A. Gonsalves, C. Toth, K. Nakamura, C. Geddes, E. Esarey, C. Schroeder, and S. Hooker, *GeV electron beams from a centimetre-scale accelerator*, Nature Physics **2** (2006) 696–699.

[52] W. P. Leemans, C. G. R. Geddes, J. Faure, C. Tóth, J. van Tilborg, C. B. Schroeder, E. Esarey, G. Fubiani, D. Auerbach, B. Marcelis, M. A. Carnahan, R. A. Kaindl, J. Byrd, and M. C. Martin, *Observation of Terahertz Emission from a Laser-Plasma Accelerated Electron Bunch Crossing a Plasma-Vacuum Boundary*, Phys. Rev. Lett. **91**(7) (2003) 074802.

[53] G. Lehmann, E. W. Laedke, and K. H. Spatschek, *Stability and evolution of one-dimensional relativistic solitons on the ion time scale*, Physics of Plasmas **13**(9) (2006) 092302.

[54] G. Lehmann, E. W. Laedke, and K. H. Spatschek, *Localized wake-field excitation and relativistic wave-breaking*, Physics of Plasmas **14**(10) (2007) 103109.

[55] B. Li, S. Ishiguro, M. Škorić, and H. Takamaru, *Stimulated Raman Scattering, Raman Cascade and Large Amplitude Relativistic Electromagnetic Solitons in Ultraintense Laser Interaction with an Underdense Plasma*, JPFRS **6** (2004) 275–278.

[56] B. Li, S. Ishiguro, M. M. Škorić, and T. Sato, *Stimulated trapped eletron acoustic wave scattering, electromagnetic soliton and ion intense interaction with subcritical plasmas*, Plasma Phys. **14** (2007) 032101.

[57] M. Lontano, S. Bulanov, and J. Koga, *One-dimensional electromagnetic solitons in a hot electron-positron plasma*, Physics of Plasmas **8**(12) (2001) 5113–5120.

[58] M. Lontano, M. Passoni, and S. Bulanov, *Relativistic electromagnetic solitons in a warm quasineutral electron-ion plasma*, Plasma Phys. **10** (2003) 639–649.

Bibliography

[59] A. Macchi, A. Bigongiari, F. Ceccherini, F. Cornolti, T. V. Liseikina, M. Borghesi, S. Kar, and L. Romagnani, *Ion dynamics and coherent structure formation following laser pulse self-channeling*, Plasma Physics and Controlled Fusion **49**(12B) (2007) B71–B78.

[60] V. G. Makhankov, *Dynamics of classical solitons (in non-integrable systems)*, Phys. Rep. (Sect. C of Phys. Lett.) **35** (1978) 1–128.

[61] V. Malka, S. Fritzler, E. Lefebvre, M.-M. Aleonard, F. Burgy, J.-P. Chambaret, J.-F. Chemin, K. Krushelnick, G. Malka, S. P. D. Mangles, Z. Najmudin, M. Pittman, J.-P. Rousseau, J.-N. Sheurer, B. Walton, and A. E. Dangor, *Electron acceleration by a wakefield forced by ab intense ultrashort laser pulse*, Science **298** (2002) 1596.

[62] J. G. Malkin, *Theorie der Stabilität einer Bewegung*, R. Oldenbourg, München, 1959.

[63] A. Mančić, L. Hadžievski, and M. Škorić, *Dynamics of electromagnetic solitons in a relativistic plasma*, Phys. Plasmas **13** (2006) 052305.

[64] M. Marklund and P. K. Shukla, *Nonlinear collective effects in photon-photon and photon-plasma interactions*, Reviews of Modern Physics **78**(2) (2006) 591.

[65] M. Martinez, E. Gaul, T. Ditmire, S. Douglas, D. Gorski, W. Henderson, J. Blakeney, D. Hammond, M. Gerity, J. Caird, A. Erlandson, I. Iovanovic, C. Ebbers, and B. Molander, *The Texas Petawatt Laser*, Laser-Induced Damage in Optical Materials: 2005 **5991**(1) (2005) 59911N.

[66] K. Mima, M. S. Jovanović, Y. Sentoku, Z.-M. Sheng, M. M. Škorić, and T. Sato, *Stimulated photon cascade and condensate in a relativistc laser-plasma interaction*, Phys. Plasmas **8**(5) (2001) 2349–2356.

Bibliography

[67] A. Modena, Z. Najmudin, A. E. Dangor, C. E. Clayton, K. A. Marsh, C. Joshi, V. Malka, C. B. Darrow, C. Danson, D. Neely, and F. N. Walsh, *Electron acceleration from the breaking of relativistic plasma waves*, Nature **377** (1995) 606–608.

[68] G. A. Mourou, T. Tajima, and S. V. Bulanov, *Optics in the relativistic regime*, Reviews of Modern Physics **78**(2) (2006) 309.

[69] N. Naumova, S. Bulanov, T. Esirkepov, F. Farina, and K. Nishihara, *Formation of Electromagnetic Postsolitons in Plasmas*, Phys. Rev. Lett. **87** (2001) 185004.

[70] M. Perry, *Crossing the Petawatt Threshold*, LLNL Science and Technology Review **4** (1996).

[71] A. S. Pirozhkov, J. Ma, M. Kando, T. Z. Esirkepov, Y. Fukuda, L.-M. Chen, I. Daito, K. Ogura, T. Homma, Y. Hayashi, H. Kotaki, A. Sagisaka, M. Mori, J. K. Koga, T. Kawachi, H. Daido, S. V. Bulanov, T. Kimura, Y. Kato, and T. Tajima, *Frequency multiplication of light back-reflected from a relativistic wake wave*, Physics of Plasmas **14**(12) (2007) 123106.

[72] S. Poornakala, A. Das, P. K. Kaw, A. Sen, Z.-M. Sheng, Y. Sentoku, K. Mima, and K. Nishkawa, *Weakly relativistic one-dimensional laser pulse envelope solitons in a warm plasma*, Plasma Phys. **9** (2002) 3802–3810.

[73] S. Poornakala, A. Das, and P. K. Sen, A.; Kaw, *Laser envelope solitons in cold overdense plasmas*, Plasma Phys. **9** (2002) 1820–1823.

[74] J. B. Rosenzweig, *Trapping, thermal effects, and wave breaking in the nonlinear plasma wake-field accelerator*, Phys. Rev. A **38**(7) (1988) 3634–3642.

[75] V. Saxena, A. Das, A. Sen, and P. Kaw, *Fluid simulation studies of the dynamical behavior of one-dimensional relativistic electromagnetic solitons*, Physics of Plasmas **13**(3) (2006) 032309.

Bibliography

[76] V. Saxena, A. Das, S. Sengupta, P. Kaw, and A. Sen, *Stability of nonlinear one-dimensional laser pulse solitons in a plasma*, Physics of Plasma **14** (2007).

[77] G. Schmidt, *Stability of Envelope Solitons*, Phys. Rev. Lett. **34**(12) (1975) 724–726.

[78] C. B. Schroeder, E. Esarey, and B. A. Shadwick, *Warm wave breaking of nonlinear plasma waves with arbitrary phase velocities*, Physical Review E **72**(5) (2005) 055401.

[79] C. B. Schroeder, E. Esarey, and B. A. Shadwick, *Comment on "Wave-breaking limits for relativistic electrostatic waves in a one-dimensional warm plasma" [Phys. Plasmas [bold 13], 123102 (2006)]*, Physics of Plasmas **14**(8) (2007) 084701.

[80] C. B. Schroeder, E. Esarey, B. A. Shadwick, and W. P. Leemans, *Trapping, dark current, and wave breaking in nonlinear plasma waves*, Physics of Plasmas **13**(3) (2006) 033103.

[81] Y. Sentoku, T. Esirkepov, K. Mima, K. Nishihara, F. Califano, F. Pegoraro, H. Sakagami, Y. Kitagawa, N. Naumova, and S. Bulanov, *Bursts of Superreflected Laser Light from Inhomogeneous Plasma due to the Generation of Relativistic Solitary Waves*, Phys. Rev. Lett. **83** (1999) 3434–3437.

[82] P. Sprangle, E. Esarey, A. Ting, and G. Joyce, *Laser wakefield acceleration and relativistic optical guiding*, Applied Physics Letters **53**(22) (1988) 2146–2148.

[83] D. Strickland and G. Mourou, *Compression of amplified chirped optical pulses*, Optics Communications **55** (1985) 447–449.

[84] T. Tajima and J. M. Dawson, *Laser electron accelerator*, Phys. Rev. Lett. **43** (1979) 267.

Bibliography

[85] A. Ting, K. Krushelnick, C. I. Moore, H. R. Burris, E. Esarey, J. Krall, and P. Sprangle, *Temporal Evolution of Self-Modulated Laser Wakefields Measured by Coherent Thomson Scattering*, Phys. Rev. Lett. **77**(27) (1996) 5377–5380.

[86] R. M. G. M. Trines and P. A. Norreys, *Wave-breaking limits for relativistic electrostatic waves in a one-dimensional warm plasma*, Phys. Plasmas **13** (2006) 123102.

[87] M. Tushentsov, A. Kim, D. Cattani, D. Anderson, and M. Lisak, *Electromagnetic Energy Penetration in the Self-Induced Transparency Regime of Relativistic Laser-Plasma Interaction*, Phys. Rev. Lett. **87** (2001) 275002.

[88] M. G. Vakhitov and A. A. Kolokolov, Radiophys. Quantum Electron. **16** (1973) 783.

VDM Verlagsservicegesellschaft mbH

Die VDM Verlagsservicegesellschaft sucht für wissenschaftliche Verlage abgeschlossene und herausragende

Dissertationen, Habilitationen, Diplomarbeiten, Master Theses, Magisterarbeiten usw.

für die kostenlose Publikation als Fachbuch.

Sie verfügen über eine Arbeit, die hohen inhaltlichen und formalen Ansprüchen genügt, und haben Interesse an einer honorarvergüteten Publikation?

Dann senden Sie bitte erste Informationen über sich und Ihre Arbeit per Email an *info@vdm-vsg.de*.

Sie erhalten kurzfristig unser Feedback!

VDM Verlagsservicegesellschaft mbH
Dudweiler Landstr. 99
D - 66123 Saarbrücken
www.vdm-vsg.de

Telefon +49 681 3720 174
Fax   +49 681 3720 1749

Die VDM Verlagsservicegesellschaft mbH vertritt

Printed by Books on Demand GmbH, Norderstedt / Germany